Springer Geophysics

The Springer Geophysics series seeks to publish a broad portfolio of scientific books, aiming at researchers, students, and everyone interested in geophysics. The series includes peer-reviewed monographs, edited volumes, textbooks, and conference proceedings. It covers the entire research area including, but not limited to, applied geophysics, computational geophysics, electrical and electromagnetic geophysics, geodesy, geodynamics, geomagnetism, gravity, lithosphere research, paleomagnetism, planetology, tectonophysics, thermal geophysics, and seismology.

More information about this series at http://www.springer.com/series/10173

Fernando Sansò · Federica Migliaccio

Quantum Measurement of Gravity for Geodesists and Geophysicists

Springer

Fernando Sansò
Milan, Italy

Federica Migliaccio
Milan, Italy

ISSN 2364-9119 ISSN 2364-9127 (electronic)
Springer Geophysics
ISBN 978-3-030-42837-2 ISBN 978-3-030-42838-9 (eBook)
https://doi.org/10.1007/978-3-030-42838-9

This Springer imprint is published by the registered company Springer Nature Switzerland AG
The registered company address is: Gewerbestrasse 11, 6330 Cham, Switzerland

Preface

During the past 30 years a great advancement in low-energy physics, particularly interactions of atoms with the electromagnetic field, has been achieved. Quoting the Nobel Prize talk of C. Cohen-Tannoudji, we can say that the development of electronics and laser techniques has allowed to implement a fine manipulation of atoms with photons. In this way, following the theory already worked out in the 50s, physicists have learnt how to cool a sample of atoms at the level of the microkelvin (μk) and, nowadays, even in the range of the nanokelvin (nk).

A wealth of important applications has sprung out from this ability of manipulating large samples ($N \sim 10^7$) of cold atoms; among them we mention, regarding the improvement of atomic clocks, the creation of atomic gyroscopes and of atomic gravity meters. This last item is obviously of great interest to geodesists and geophysicists, particularly for potential applications to space geodesy.

More than one year ago my co-author, Federica Migliaccio, asked me to look into the matter and give a seminar to a group of scientists of the Italian geodetic/geophysical community. After some discussion we soon realized that the initiative could be carried out following one of two completely different approaches: either to provide a "description" of the phenomena that permit the quantum measurement of gravity, or to "explain" the physical principles that are at the basis of what we could call with R. P. Feynman a theoretical experiment.

In fact, many geodesists and geophysicists do not have the scientific background to thoroughly understand the quantum process leading to the measurement of g, so we decided to follow the second approach, also in view of the many different future applications of such matters in our disciplines; be it enough to mention the direct measurement of gravity potential differences by the observation of the beat of different clocks.

True is that one can always claim that there is some magic box producing numbers (measurements) and we should only be interested in how to use them for our purposes, however we firmly believe that completely losing contact with the science behind the measurement can lead us in the long run into a dead-end street.

This is the why this text was born, conceived as a road that leads the reader from classical physics (mechanics and electromagnetism, considered as a common scientific background), to the basics of quantum mechanics, and finally to understand the dynamics of a bunch of atoms launched through the gravity field, while interacting with suitably resonant laser beams.

Of course, this is not a general text on theoretical physics, since it only explains those concepts that are necessary to understand atom manipulation by photons. Among these, however, there are the principles of quantum mechanics, presented in a special chapter, following my own views, from the period of my studies in physics.

We notice as well that there are basically two types of measurements of g, one based on stimulated Raman transition, and the other on so-called Block oscillations, although a new generation of instruments based on Bose–Einstein condensation is currently under study.

Among them, we have chosen to present the former, which can be easily worked out analytically. However, the principles presented in the text can provide a good starting point to understand also the latter.

The reader will realize that in the References most textbooks are in Italian. This is because these are the books on which I studied during my university time, yet the reader will realize too that most of them are famous books from the founders of quantum mechanics and, though mentioned in an Italian translation, they say the same things as in the original language.

Milan, Italy
October 2019

Fernando Sansò

Contents

1 Recalls of Classical Mechanics 1
1.1 Introduction 1
1.2 From Newton to Lagrange 1
1.3 From Lagrange to Hamilton 4
1.4 Poisson Parentheses 6
1.5 Canonical Transformations. Conjugate Variables 8
1.6 Dynamic Systems with Generalized Potentials 11
References 14

2 Recalls of the Classical Theory of the Electromagnetic Field 15
2.1 Introduction 15
2.2 Maxwell Equations and Lorentz Equation 15
2.3 Electromagnetic Waves in Empty Space 17
2.4 Wave Packets 22
2.5 Electromagnetic Potentials, Lorentz Gauge 24
2.6 Energy Conservation 31
2.7 Momentum Conservation 34
References 39

3 The Crisis of Classical Physics and the Dawn of Quantum Physics 41
3.1 Introduction 41
3.2 The Untenable Planetary Model of the Atom 42
3.3 The Birth of Quanta 45
3.4 The Electromagnetic Radiation Behaves Like a Flow of Particles 46
3.5 The Quantized Atom of Niels Bohr 49
3.6 Overcoming the Dualism Waves–Particles. The "Waves of Matter" of Louis de Broglie 51

3.7 The Heisenberg Indetermination Principle 54
References 57

4 The Principles of Quantum Mechanics 59
4.1 Introduction 59
4.2 A Theorem on Fourier Transform 60
4.3 Some Facts About Hermitian Operators in Hilbert Spaces 65
4.4 A More Precise Formulation of Seven Principles of Quantum Mechanics 71
4.5 The Time Evolution of a Quantum State: Schrödinger Equation 77
4.6 Stationary States; Constants of Motion 80
References 83

5 First Applications of Quantum Theory 85
5.1 Introduction 85
5.2 The Free Particle 85
5.3 The Hydrogen Atom as an Example 87
5.4 The Quantum Theory of Angular Momentum 91
5.5 First-Order Spectral Analysis of Static Perturbations of the Hamiltonian 93
5.6 The Zeeman Effect, the Spin, and the Atomic Spectrum 96
5.7 Time-Dependent Perturbations of the Hamiltonian 101
References 104

6 The Quantum Measurement of Gravity 105
6.1 Introduction 105
6.2 The Dipole Approximation of the Optical Radiation—Atom Interaction 106
6.3 Rabi Equation 111
6.4 Adding Momentum Exchange to the Atom–Photon Interaction 117
6.5 The Atom Interferometry and the Measurement of $\mathbf{g}$ 123
6.6 Complements and Improvements 126
6.7 Results and Perspectives 129
References 132

Chapter 1
Recalls of Classical Mechanics

1.1 Introduction

The chapter starts from common knowledge of physics, namely Newton's equation, briefly recalling the Lagrangian and Hamiltonian formalism, as well as the Hamilton principle of least action.

We pass then to the definition and properties of Poisson brackets, with the purpose of defining canonical variables and transformations. Of fundamental importance for the axiomatic construction of quantum theory, is the result that energy and time are a canonical couple.

The set up of Hamiltonian theory for generalized potentials, depending also on velocities, will be important too for the purposes of this text.

The argument is very classical, and the literature on it is immense. A typical citation would be [1], although this chapter has been mainly constucted by using [2, 3].

1.2 From Newton to Lagrange

We concentrate on the dynamics of a single particle in an external field of forces $\boldsymbol{F}(\boldsymbol{x})$. We assume $\boldsymbol{F}$ to be a potential field and, contrary to the geodetic tradition, we put

$$\boldsymbol{F} = -\nabla U \ , \tag{1.1}$$

in this way U has the meaning of a potential energy, as we shall see later. The dynamics of the particle is controlled by Newton's equation

$$m\ddot{\boldsymbol{x}} = -\nabla U(\boldsymbol{x}) \tag{1.2}$$

or, defining the linear momentum

F. Sansò and F. Migliaccio, *Quantum Measurement of Gravity for Geodesists and Geophysicists*, Springer Geophysics,
https://doi.org/10.1007/978-3-030-42838-9_1

$$\boldsymbol{p} = m\dot{\boldsymbol{x}} \,, \tag{1.3}$$

$$\dot{\boldsymbol{p}} = -\nabla U(\boldsymbol{x}) \,. \tag{1.4}$$

We ask ourselves whether (1.2) can be derived from a variational principle, more precisely we ask ourselves whether the physical motion $\boldsymbol{x}(t)$ between two points

$$\boldsymbol{x}_1 = \boldsymbol{x}(t_1) \,, \; \boldsymbol{x}_2 = \boldsymbol{x}(t_2) \tag{1.5}$$

where $(\boldsymbol{x}_1, \boldsymbol{x}_2, t_1, t_2)$ are fixed, has the property to make some functional $S[\boldsymbol{x}(t)]$ stationary with respect to other, not physical, motions that respect (1.5).

Theorem 1.1 *Considering the Lagrangian function of the system,*

$$L(\boldsymbol{x}, \dot{\boldsymbol{x}}) = \frac{m}{2}|\dot{\boldsymbol{x}}|^2 - U(\boldsymbol{x}) \,, \tag{1.6}$$

the functional S*, called the action of the system between* t_1 *and* t_2 *when computed in correspondence to the true motion, given by*

$$S[\boldsymbol{x}(t)] = \int_{t_1}^{t_2} L(\boldsymbol{x}, \dot{\boldsymbol{x}}) dt \tag{1.7}$$

is stationary with respect to variations of $\boldsymbol{x}(t)$.

The corresponding variational principle is called least action *or* Hamilton principle.

Proof We prove that the condition of stationarity of (1.7) gives (1.2). In fact,

$$\begin{aligned} \delta S &= \int_{t_1}^{t_2} L(\boldsymbol{x} + \delta\boldsymbol{x} \,,\; \dot{\boldsymbol{x}} + \delta\dot{\boldsymbol{x}}) dt = \\ &= (\text{first order in } \delta\boldsymbol{x}) = \int_{t_1}^{t_2} \left[\frac{\partial L}{\partial \boldsymbol{x}} \delta\boldsymbol{x} + \frac{\delta L}{\partial \dot{\boldsymbol{x}}} \delta\dot{\boldsymbol{x}} \right] dt. \end{aligned} \tag{1.8}$$

But

$$\begin{aligned} \frac{\partial L}{\partial \boldsymbol{x}} &= -\nabla U(\boldsymbol{x})^T = \boldsymbol{F}(\boldsymbol{x})^T \\ \frac{\partial L}{\partial \dot{\boldsymbol{x}}} &= m\dot{\boldsymbol{x}}^T = \boldsymbol{p}^T \end{aligned}$$

and, recalling that $\delta\boldsymbol{x}(t_1) = \delta\boldsymbol{x}(t_2) = 0$,

$$\int_{t_1}^{t_2} \frac{\partial L}{\partial \dot{\boldsymbol{x}}} \delta\dot{\boldsymbol{x}} \, dt = \left[\frac{\partial L}{\partial \dot{\boldsymbol{x}}} \delta\boldsymbol{x} \right]_{t_1}^{t_2} - \int_{t_1}^{t_2} \frac{d}{dt} \frac{\partial L}{\partial \dot{\boldsymbol{x}}} \delta\boldsymbol{x} \, dt$$

so that (1.8) becomes

$$\delta S = \int_{t_1}^{t_2} [\boldsymbol{F} - \frac{d}{dt}\boldsymbol{p}]^T \delta \boldsymbol{x} \, dt = 0 \qquad \forall \delta \boldsymbol{x}$$

i.e.

$$F = \frac{d}{dt}\boldsymbol{p} .$$

□

The function of the velocity

$$T = \frac{m}{2}|\dot{\boldsymbol{x}}|^2 = \frac{1}{2m}p^2 \tag{1.9}$$

is the kinetic energy, so that the Lagrangian of the system is generally written as

$$L(\boldsymbol{x}, \dot{\boldsymbol{x}}) = T - U . \tag{1.10}$$

Let us observe that L has the dimension of an energy but it is not an energy because it is not conserved in time, in fact a direct computation shows that

$$\begin{aligned} \frac{d}{dt}L &= \frac{\partial L}{\partial \boldsymbol{x}}\dot{\boldsymbol{x}} + \frac{\partial L}{\partial \dot{\boldsymbol{x}}}\ddot{\boldsymbol{x}} = \boldsymbol{F} \cdot \dot{\boldsymbol{x}} + m\dot{\boldsymbol{x}} \cdot \ddot{\boldsymbol{x}} \\ &= 2\boldsymbol{F} \cdot \dot{\boldsymbol{x}} \neq 0 \quad \text{(in general)}. \end{aligned}$$

Remark 1.1 Let us further notice that, if we change coordinates from Cartesian, $\boldsymbol{x}$, to a general set $\boldsymbol{q}$, the kinetic energy takes a form depending on $\boldsymbol{q}$, but still quadratic in $\dot{\boldsymbol{q}}$, i.e.

$$T = \frac{m}{2}|\dot{x}|^2 = \frac{m}{2}\dot{\boldsymbol{q}}^T Q(\boldsymbol{q})\boldsymbol{q} \tag{1.11}$$

with $Q(\boldsymbol{q})$ the metric tensor in the $\boldsymbol{q}$ coordinates.

In any case, the dynamics is always defined by the Hamilton principle and the corresponding Lagrange equations

$$\frac{d}{dt}\frac{\partial L}{\partial \dot{\boldsymbol{q}}} - \frac{\partial L}{\partial \boldsymbol{q}} = 0 \tag{1.12}$$

still hold.

The vector

$$p = \nabla_{\dot{q}} L = \left(\frac{\partial L}{\partial \dot{q}}\right)^T \tag{1.13}$$

is called *generalized kinematic momentum* and

$$\mathscr{F} = \left(\frac{\partial L}{\partial q}\right)^T = \nabla_q L \tag{1.14}$$

is called *Lagrangian force field.*

Remark 1.2 (*non-uniqueness of the Lagrangian*) Two Lagrangians L, L' can be equivalent in describing the dynamics of a system, if

$$L' - L = \frac{d}{dt} F(x, t) . \tag{1.15}$$

In this case in fact

$$\int_{t_1}^{t_2} L' dt = \int_{t_1}^{t_2} L dt + F(x_2, t_2) - F(x_1, t_1)$$

but the last two terms are constant and their variation δF is zero.

1.3 From Lagrange to Hamilton

We start from energy conservation and the definition of the Hamilton function.

Energy conservation. From Eq. (1.12) we can write

$$\begin{aligned} D_t L &= \frac{\partial L}{\partial q} \cdot \dot{q} + \frac{\partial L}{\partial \dot{q}} \cdot \ddot{q} = \frac{\partial L}{\partial q} \cdot \dot{q} + D_t \left(\frac{\partial L}{\partial q} \cdot \dot{q}\right) - D_t \left(\frac{\partial L}{\partial \dot{q}}\right) \cdot \dot{q} \\ &\equiv D_t \left(\frac{\partial L}{\partial \dot{q}} \cdot \dot{q}\right) \equiv D_t (p \cdot \dot{q}), \end{aligned}$$

i.e.

$$D_t (p \cdot \dot{q} - L) = 0 . \tag{1.16}$$

This shows that

$$\mathscr{H} = p \cdot \dot{q} - L \tag{1.17}$$

is a quantity conserved along the motion. $\mathscr{H}$ is by definition the Hamiltonian of the system and we show that it has the physical meaning of energy of the system.

In fact

$$\boldsymbol{p}\cdot\dot{\boldsymbol{q}}=\frac{\partial L}{\partial\dot{\boldsymbol{q}}}\cdot\dot{\boldsymbol{q}}=\frac{\partial T}{\partial\dot{\boldsymbol{q}}}\cdot\dot{\boldsymbol{q}}=2T, \tag{1.18}$$

because T is homogeneous of the second degree in $\dot{\boldsymbol{q}}$, according to Euler's theorem (see (1.11)).

So

$$\mathscr{H}=2T-(T-U)=T+U\ . \tag{1.19}$$

This also clarifies that the total energy of the system is the sum of the kinetic energy and the potential energy. When one increases along the motion, the other decreases and vice versa, keeping the value of $\mathscr{H}$ constant.

Now we want to transform the description of the dynamics from the couple $(\boldsymbol{q},\dot{\boldsymbol{q}})$ to the couple $(\boldsymbol{q},\boldsymbol{p})$ by inverting the relation

$$\boldsymbol{p}=\frac{\partial L}{\partial\dot{\boldsymbol{q}}}\Rightarrow\dot{\boldsymbol{q}}=\dot{\boldsymbol{q}}(\boldsymbol{p},\boldsymbol{q})\ ;\ \frac{\partial L}{\partial\dot{\boldsymbol{q}}}[\boldsymbol{q},\dot{\boldsymbol{q}}(\boldsymbol{p},\boldsymbol{q})]\equiv\boldsymbol{p}\ . \tag{1.20}$$

From (1.17) we can write, recalling (1.12) and (1.13),

$$\begin{aligned} d\mathscr{H}&\equiv d\boldsymbol{p}\cdot\dot{\boldsymbol{q}}+\boldsymbol{p}\cdot d\dot{\boldsymbol{q}}-\frac{\partial L}{\partial\boldsymbol{q}}\cdot d\boldsymbol{q}-\frac{\partial L}{\partial\dot{\boldsymbol{q}}}\cdot d\dot{\boldsymbol{q}}=\\ &=d\boldsymbol{p}\cdot\dot{\boldsymbol{q}}-\dot{\boldsymbol{p}}\cdot d\boldsymbol{q}\ ; \end{aligned} \tag{1.21}$$

such equation implies

$$\dot{\boldsymbol{q}}=\frac{\partial\mathscr{H}}{\partial\boldsymbol{p}} \tag{1.22}$$

$$\dot{\boldsymbol{p}}=-\frac{\partial\mathscr{H}}{\partial\boldsymbol{q}}\ . \tag{1.23}$$

So we have succeeded in transforming the equations of motion from a second-order differential (vector) equation to a system of two differential equations of the first order, possessing the particular anti-symmetry (1.22), (1.23). This is known in mathematical physics as *symplectic structure*. These equations are known as Hamilton equations and they are equivalent to Lagrange equation in describing the dynamics of the system. Of course also Hamilton equations can be derived from a variation of the action functional S, because

$$S=\int_{t_1}^{t_2}Ldt\equiv\int_{t_1}^{t_2}(\boldsymbol{p}\cdot\dot{\boldsymbol{q}}-\mathscr{H})dt\ . \tag{1.24}$$

Equations (1.22), (1.23) are obtained from the variation of S equal to zero, considering $\boldsymbol{q}(t)$, $\boldsymbol{p}(t)$ as independent variables and imposing on $\boldsymbol{q}(t)$ only, the constraints $\delta\boldsymbol{q}(t_1) = \delta\boldsymbol{q}(t_2) = 0$.

Remark 1.3 It might happen that L is explicitly depending on time t, e.g. because $U = U(\boldsymbol{x}, t)$. This is what happens, for instance, in satellite geodesy when the gravitational potential (here changed of sign) is seen in an inertial (non-rotating with the Earth) reference system.

In this case we have

$$\begin{aligned} D_t L(\boldsymbol{q}, \dot{\boldsymbol{q}}, t) &= \frac{\partial L}{\partial \boldsymbol{q}} \cdot \dot{\boldsymbol{q}} + \frac{\partial L}{\partial \dot{\boldsymbol{q}}} \cdot \ddot{\boldsymbol{q}} + \frac{\partial L}{\partial t} \\ &= \dot{\boldsymbol{p}} \cdot \dot{\boldsymbol{q}} + \boldsymbol{p} \cdot \ddot{\boldsymbol{q}} + \frac{\partial L}{\partial t} = D_t(\boldsymbol{p} \cdot \dot{\boldsymbol{q}}) + \frac{\partial L}{\partial t} \end{aligned} \tag{1.25}$$

or, recalling (1.11),

$$D_t(\mathscr{H}) = -\frac{\partial L}{\partial t} \left(= \frac{\partial U}{\partial t} \right) . \tag{1.26}$$

Furthermore,

$$\begin{aligned} D_t(\mathscr{H}) &= \frac{\partial \mathscr{H}}{\partial \boldsymbol{q}} \cdot \dot{\boldsymbol{q}} + \frac{\partial \mathscr{H}}{\partial \boldsymbol{p}} \cdot \dot{\boldsymbol{p}} + \frac{\partial \mathscr{H}}{\partial t} \equiv \\ &\equiv -\dot{\boldsymbol{p}} \cdot \dot{\boldsymbol{q}} + \dot{\boldsymbol{q}} \cdot \dot{\boldsymbol{p}} + \frac{\partial \mathscr{H}}{\partial t} \equiv \frac{\partial \mathscr{H}}{\partial t} , \end{aligned} \tag{1.27}$$

meaning that energy is not anymore conserved.

1.4 Poisson Parentheses

Given two functions of the state variables of a dynamic system $(\boldsymbol{q}, \boldsymbol{p})$, $f(\boldsymbol{q}, \boldsymbol{p}, t)$, $g(\boldsymbol{q}, \boldsymbol{p}, t)$, we define the Poisson parentheses of f and g as

$$\{f, g\} \equiv \frac{\partial f}{\partial \boldsymbol{p}} \cdot \frac{\partial g}{\partial \boldsymbol{q}} - \frac{\partial f}{\partial \boldsymbol{q}} \cdot \frac{\partial g}{\partial \boldsymbol{p}} . \tag{1.28}$$

Lemma 1.1 (Poisson Lemma) *For any dynamic system, characterized by the Hamiltonian $\mathscr{H}$, the Poisson parenthesis of any function of the state variables, $f = f(\boldsymbol{q}, \boldsymbol{p}, t)$, and $\mathscr{H}$, determines the evolution of f according to*

$$D_t f \equiv \frac{\partial f}{\partial t} + \{\mathscr{H}, f\} . \tag{1.29}$$

Proof We have

$$
\begin{aligned}
D_t f &= \frac{\partial f}{\partial \boldsymbol{q}} \cdot \dot{\boldsymbol{q}} + \frac{\partial f}{\partial \boldsymbol{p}} \cdot \dot{\boldsymbol{p}} + \frac{\partial f}{\partial t} \equiv \\
&\equiv \frac{\partial f}{\partial \boldsymbol{q}} \cdot \frac{\partial \mathscr{H}}{\partial \boldsymbol{p}} - \frac{\partial f}{\partial \boldsymbol{p}} \cdot \frac{\partial \mathscr{H}}{\partial \boldsymbol{q}} + \frac{\partial f}{\partial t} ,
\end{aligned}
$$

whence (1.29). □

Corollary 1.1 *Any constant of the motion f (namely a quantity that preserves its value in time) satisfies the equation*

$$
\frac{\partial f}{\partial t} + \{\mathscr{H}, f\} = 0 ; \tag{1.30}
$$

in particular, if f does not depend on t explicitly, f is conserved if and only if

$$
\{\mathscr{H}, f\} = 0 . \tag{1.31}
$$

Example 1.1 For future use we compute the Poisson parentheses of the components q_i, p_k. One has

$$
\{q_i, q_k\} = 0 , \; \{p_i, p_k\} = 0 , \; \{p_i, q_k\} = \delta_{ik} . \tag{1.32}
$$

Example 1.2 We note as well that the following formulas hold for any $f = f(\boldsymbol{q}, \cdot \boldsymbol{p})$:

$$
\{f, p_i\} = -\frac{\partial f}{\partial q_i} \tag{1.33}
$$

$$
\{f, q_k\} = \frac{\partial f}{\partial p_k} . \tag{1.34}
$$

So, if we take $f = \mathscr{H}(\boldsymbol{q}, \boldsymbol{p})$ and we apply (1.30), we can see that (1.33), (1.34) become again the Hamilton equations.

Example 1.3 Let our system be that of a free particle, i.e.

$$
\mathscr{H} = \frac{1}{2m} p^2 ,
$$

then $\boldsymbol{p}$ is conserved. In fact

$$
\{\mathscr{H}, p_i\} = \frac{1}{2m}\{p^2, p_i\} = \frac{1}{2m} \sum_k \left(\frac{\partial p^2}{\partial p_k} \frac{\partial p_i}{\partial q_k} - \frac{\partial p^2}{\partial q_k} \frac{\partial p_i}{\partial p_k} \right) \equiv 0 ,
$$

because $\frac{\partial p_i}{\partial q_k} = 0$ and $\frac{\partial p^2}{\partial q_k} = 0$.

In a similar way one can prove that for a particle in a central field, i.e. $U = U(r)$, the angular momentum

$$\boldsymbol{M} = \boldsymbol{x} \wedge \boldsymbol{p}$$

is conserved.

1.5 Canonical Transformations. Conjugate Variables

Let us consider a dynamic system with Hamiltonian $\mathscr{H}$ and state variables $(\boldsymbol{q}, \boldsymbol{p})$, that evolve according to Hamilton equations

$$\dot{\boldsymbol{q}} = \frac{\partial \mathscr{H}}{\partial \boldsymbol{p}} , \; \dot{\boldsymbol{p}} = -\frac{\partial \mathscr{H}}{\partial \boldsymbol{q}} , \tag{1.35}$$

which in turn descend from the stationarity of the action S,

$$S \equiv \int_{t_1}^{t_2} (\boldsymbol{p} \cdot \dot{\boldsymbol{q}} - \mathscr{H}) dt . \tag{1.36}$$

It is obvious that we can always change our system variables to

$$\boldsymbol{Q} = \boldsymbol{Q}(\boldsymbol{q}, \boldsymbol{p}, t) , \; \boldsymbol{P} = \boldsymbol{P}(\boldsymbol{q}, \boldsymbol{p}, t) \tag{1.37}$$

and, when (1.37) are invertible, we find the way to define the dynamics of the system by using the Poisson parentheses and Eq. (1.29).

However, in general, it does not happen that the new equations have the same Hamiltonian form, of course with a new Hamiltonian $\mathscr{H}'(\boldsymbol{Q}, \boldsymbol{P})$.

When this happens, we can as well define a new action S' by

$$S' \equiv \int_{t_1}^{t_2} \{\boldsymbol{P} \cdot \dot{\boldsymbol{Q}} - \mathscr{H}'\} dt . \tag{1.38}$$

Now the stationarity of S or S' are equivalent if

$$S' - S = \text{const} , \tag{1.39}$$

which, as already observed for Lagrange equations (see Remark 1.2), happens only if the two integrands in (1.36) and (1.38) differ by a total derivative of a function with respect to time, i.e.

$$\boldsymbol{p} \cdot \dot{\boldsymbol{q}} - \mathscr{H} = \boldsymbol{P} \cdot \dot{\boldsymbol{Q}} - \mathscr{H}' + D_t F . \tag{1.40}$$

The function F is called the generating function of the canonical transformation; here it is convenient to choose F as function of the couple $(\boldsymbol{q}, \boldsymbol{Q})$. In fact, it is clear that, by exploiting the first of (1.37), we can derive $\boldsymbol{p} = \boldsymbol{p}(\boldsymbol{q}, \boldsymbol{Q}, t)$, that substituted into the second of (1.37) gives $\boldsymbol{P} = \boldsymbol{P}(\boldsymbol{q}, \boldsymbol{Q}, t)$. With such a choice we can write

$$D_t F(q, Q, t) = F_{\boldsymbol{q}} \cdot \dot{\boldsymbol{q}} + F_{\boldsymbol{Q}} \cdot \dot{\boldsymbol{Q}} + \frac{\partial F}{\partial t} . \tag{1.41}$$

Comparing (1.40) and (1.41) we can see that

$$\boldsymbol{p} = F_{\boldsymbol{q}} \tag{1.42}$$

$$\boldsymbol{P} = -F_{\boldsymbol{Q}} \tag{1.43}$$

$$\mathcal{H}' = \mathcal{H} + \frac{\partial F}{\partial t} . \tag{1.44}$$

These equations completely define the canonical transformation.

The following theorem expresses a remarkable relation between Poisson parentheses and canonical transformations.

Theorem 1.2 *The Poisson equations are intrinsic to the Hamiltonian dynamics of the system in the sense that they are invariant under canonical transformations.*

Proof Remember that $(\boldsymbol{q}, \boldsymbol{p})$ are our old variables, while $(\boldsymbol{Q}, \boldsymbol{P})$ are our new variables. To simplify the proof, assume that q, p are one-dimensional.

From (1.42) we derive

$$p_Q = F_{qQ} + F_{qq}\, q_Q$$
$$p_P = F_{qq}\, q_P \, ;$$

from (1.43) we get

$$0 = -F_{Qq}\, q_Q - F_{QQ}$$
$$1 = -F_{Qq}\, q_P \, .$$

Combining the four equations above, we find

$$\begin{cases} q_Q = -\frac{F_{QQ}}{F_{Qq}} \\ q_P = -\frac{1}{F_{Qq}}, \end{cases} \tag{1.45}$$

$$\begin{cases} p_Q = -\frac{1}{F_{Qq}}(F_{qq}F_{QQ} - F_{Qq}^2) \\ p_P = -\frac{F_{qq}}{F_{Qq}}. \end{cases} \tag{1.46}$$

With the help of (1.45), (1.46), for any given $f = f(q, p, t)$ we can write

$$f_Q = f_q q_Q + f_p p_Q = -f_q \frac{F_{QQ}}{F_{Qq}} - f_p \frac{1}{F_{Qq}}(F_{qq}F_{QQ} - F_{Qq}^2)$$
$$f_P = f_q q_P + f_p p_P = -f_q \frac{1}{F_{Qq}} - f_p \frac{F_{qq}}{F_{Qq}}.$$

By using these relations, it is only a matter of algebra to compute

$$\{f, g\}_{QP} = \frac{\partial f}{\partial P}\frac{\partial g}{\partial Q} - \frac{\partial f}{\partial Q}\frac{\partial g}{\partial p}$$

and verify that

$$\{f, g\}_{QP} \equiv \{f, g\}_{qp} \,. \tag{1.47}$$

The vector case is a straightforward generalization. □

Definition of Conjugate Variables. Any couple (Q, P) derived from (q, p) by a canonical transformation defines two conjugate variables.

Corollary 1.2 (of Theorem 1.2) *Any couple of conjugate vector variables satisfies the relations*

$$\{Q_i, Q_k\} = 0\,,\ \{P_i, P_k\} = 0\,,\ \{Q_i, P_k\} = \delta_{ik}\,. \tag{1.48}$$

Example 1.4 In this example we show that, at least for a system in R^1 with the ordinary Hamiltonian, in Cartesian coordinates,

$$H = \frac{1}{2m}p^2 + V(q)\,, \tag{1.49}$$

one can take $Q = H$ and $P \equiv -t$ as a couple of conjugate variables.

In fact, let us start noticing that from (1.49) one has

$$p = \sqrt{2m(H - V)}\,, \tag{1.50}$$

implying that, if we define the generating function

$$F(q, Q) = \int_{q_0}^{q} \sqrt{2m(Q - V(q'))}dq' \tag{1.51}$$

the relation (1.42), namely

$$p \equiv \frac{\partial F}{\partial q} = \sqrt{2m(Q - V(q))}$$

is obviously satisfied.

Then, according to (1.43), one has

$$P = -\frac{\partial F}{\partial Q} \equiv -\sqrt{m}\int_{q_0}^{q} \frac{1}{\sqrt{2(Q-V)}} dq' \,. \tag{1.52}$$

On the other hand, referring to the very definition $p = m\frac{dq}{dt}$, from (1.50) we can derive (recall that $H \equiv Q$)

$$\frac{dq}{dt} = \frac{1}{\sqrt{m}}\sqrt{2(Q-V)} \,; \tag{1.53}$$

inverting and integrating along the trajectory from q_0 to q we get

$$t = \sqrt{m}\int_{q_0}^{q} \frac{1}{\sqrt{2(Q-V)}} dq' \tag{1.54}$$

that, compared to (1.52), shows that

$$P \equiv -t \tag{1.55}$$

as we wanted to prove.

It is maybe worth mentioning that formula (1.54), with $Q = H$, is nothing but the integral of Newton's equation, as discovered by Maupertuis, when the energy of the system is considered as a known constant of the motion.

1.6 Dynamic Systems with Generalized Potentials

Let us consider a dynamic system (in particular a single particle) characterized by a Lagrangian that includes a generalized potential U

$$U = U(\boldsymbol{x}, \dot{\boldsymbol{x}}, t) \,, \tag{1.56}$$

we *assume* that Hamilton's principle, namely stationarity of the action (1.7), holds and we would like to know what kind of equations control the dynamics.

So we have, in Cartesian coordinates,

$$L = \frac{1}{2}m|\dot{\boldsymbol{x}}|^2 - U \tag{1.57}$$

and

$$S \equiv \int_{t_1}^{t_2} \left\{ \frac{1}{2}m|\dot{\boldsymbol{x}}|^2 - U(\boldsymbol{x}, \dot{\boldsymbol{x}}, t) \right\} dt \,. \tag{1.58}$$

The variation of (1.58) gives

$$\delta S = \int_{t_1}^{t_2} \left\{ m\dot{\boldsymbol{x}} \cdot \delta\dot{\boldsymbol{x}} - \frac{\partial U}{\partial \boldsymbol{x}} \cdot \delta\boldsymbol{x} - \frac{\partial U}{\partial \dot{\boldsymbol{x}}} \cdot \delta\dot{\boldsymbol{x}} \right\} dt\,.$$

An integration by parts, recalling that $\delta\boldsymbol{x}(t_1) = \delta\boldsymbol{x}(t_2) = 0$, gives

$$\delta S \equiv \int_{t_1}^{t_2} \left\{ -m\ddot{\boldsymbol{x}} - \frac{\partial U}{\partial \boldsymbol{x}} + D_t \frac{\partial U}{\partial \dot{\boldsymbol{x}}} \right\} \cdot \delta\boldsymbol{x}\, dt = 0$$

and consequently the Lagrange equations become

$$m\ddot{\boldsymbol{x}} = -\frac{\partial U}{\partial \boldsymbol{x}} + D_t \frac{\partial U}{\partial \dot{\boldsymbol{x}}}\,. \tag{1.59}$$

Example 1.5 (*The Schwarzschild potential*) We apply (1.59) to the case that U is linear in $\dot{\boldsymbol{x}}$, namely

$$U = V - \dot{\boldsymbol{x}} \cdot \boldsymbol{A}(\boldsymbol{x}, t)\,. \tag{1.60}$$

In this case the force field $\boldsymbol{F}$ is given by

$$\boldsymbol{F} = -\frac{\partial V}{\partial \boldsymbol{x}} + \nabla_x[\dot{\boldsymbol{x}} \cdot \boldsymbol{A}(\boldsymbol{x}, t)] - D_t\, \boldsymbol{A}(\boldsymbol{x}, t) \tag{1.61}$$

But

$$D_t\, \boldsymbol{A}(\boldsymbol{x}, t) = (\dot{\boldsymbol{x}} \cdot \nabla_x)\, \boldsymbol{A}(\boldsymbol{x}, t) + \frac{\partial \boldsymbol{A}}{\partial t}\,, \tag{1.62}$$

so that (1.61) becomes

$$\boldsymbol{F} = -\nabla_x V + \nabla_x[\dot{\boldsymbol{x}} \cdot \boldsymbol{A}(\boldsymbol{x}, t)] - (\dot{\boldsymbol{x}} \cdot \nabla_x)\, \boldsymbol{A}(\boldsymbol{x}, t) - \frac{\partial A}{\partial t}\,. \tag{1.63}$$

On the other hand, by applying the identity

$$\boldsymbol{a} \wedge (\boldsymbol{b} \wedge \boldsymbol{c}) = \boldsymbol{b}(\boldsymbol{a} \cdot \boldsymbol{c}) - (\boldsymbol{a} \cdot \boldsymbol{b})\boldsymbol{c}$$

we see that the following differential identity holds:

$$\dot{\boldsymbol{x}} \wedge (\nabla_x \wedge \boldsymbol{A}) \equiv \nabla_x(\dot{\boldsymbol{x}} \cdot \boldsymbol{A}) - (\dot{\boldsymbol{x}} \cdot \nabla_x)\boldsymbol{A}\,,$$

so that (1.63) reads

$$\boldsymbol{F} = -\nabla_x V - \frac{\partial \boldsymbol{A}}{\partial t} + \dot{\boldsymbol{x}} \wedge (\nabla_x \wedge \boldsymbol{A}) \,. \tag{1.64}$$

If we put

$$\begin{aligned} -\ \nabla_x V - \frac{\partial \boldsymbol{A}}{\partial t} &= e\boldsymbol{E} \\ \nabla_x \wedge \boldsymbol{A} &= \frac{e}{c}\,\boldsymbol{H} \end{aligned}$$

we realize that the dynamic equation (1.61) describes the motion of a particle with charge e in the e.m. field $(\boldsymbol{E}, \boldsymbol{H})$, under the Lorentz force

$$\boldsymbol{F} = e(\boldsymbol{E} + \frac{1}{c}\dot{\boldsymbol{x}} \wedge \boldsymbol{H}) \,, \tag{1.65}$$

where c denotes as customary the velocity of the light in vacuum.

To conclude the example, we want to derive the form of the Hamiltonian dynamics of this system. In particular, we need, after putting $\boldsymbol{q} = \boldsymbol{x}$, the associated moment vector $\boldsymbol{p}$ and the corresponding Hamiltonian function $\mathcal{H}$.

Following the definition (1.13) we have

$$\boldsymbol{p} = \frac{\partial L}{\partial \dot{\boldsymbol{x}}} = m\dot{\boldsymbol{x}} + \boldsymbol{A} \,, \tag{1.66}$$

i.e.

$$\dot{\boldsymbol{x}} = \frac{1}{m}(\boldsymbol{p} - \boldsymbol{A}) \tag{1.67}$$

Therefore, using alternatively (1.66) and (1.67), we get

$$\begin{aligned} \mathcal{H} &= \boldsymbol{p} \cdot \dot{\boldsymbol{x}} - L \equiv \boldsymbol{p} \cdot \dot{\boldsymbol{x}} - \frac{1}{2} m |\dot{\boldsymbol{x}}|^2 + V - \dot{\boldsymbol{x}} \cdot \boldsymbol{A} \\ &\equiv m|\dot{\boldsymbol{x}}|^2 + \dot{\boldsymbol{x}} \cdot \boldsymbol{A} - \frac{1}{2} m |\dot{\boldsymbol{x}}|^2 + V - \dot{\boldsymbol{x}} \cdot \boldsymbol{A} = \\ &= \frac{1}{2} m |\dot{\boldsymbol{x}}|^2 + V \equiv \frac{1}{2m} |\boldsymbol{p} - \boldsymbol{A}|^2 + V \equiv \\ &= \frac{1}{2m} p^2 - \frac{1}{m} \boldsymbol{p} \cdot \boldsymbol{A} + \frac{1}{2m} A^2 + V \,. \end{aligned} \tag{1.68}$$

References

1. Arnold VI (1978) Mathematical methods of classical mechanics. Springer Verlag, New York Heidelberg Berlin
2. Caldirola P, Cirelli R, Prosperi GM (1982) Introduzione alla fisica teorica. UTET, Torino
3. Landau LD, Lifšic EM, (1965) Meccanica. Boringhieri, Torino

Chapter 2
Recalls of the Classical Theory of the Electromagnetic Field

2.1 Introduction

The theory of the propagation of an electromagnetic (e.m.) field will be central for our subject that treats essentially the interaction of this field with atoms. As we will see in Chap. 6, the description of a laser beam will be classical, namely based on Maxwell equations. On the other hand, we will need as well the formulation of the theory in terms of vector and scalar potentials, because they are naturally related to the Hamiltonian of a charged particle in the field.

Also the concepts of energy and momentum of the e.m. field will be important to understand the transition from classical to quantum physics, so we will formulate and prove the corresponding conservation theorems.

Once more, the literature on the Electromagnetic field is extensive. For instance, classical texts could be [1] or [2]. In designing the chapter, though, [3] and [4] have been mostly used.

2.2 Maxwell Equations and Lorentz Equation

We are interested in the interaction between the electromagnetic (e.m.) field and charged particles because this is our ultimate goal, when we come to describe the physics of particles by quantum theory. In a classical setting the generation of the e.m. field $(\boldsymbol{E}, \boldsymbol{H})$ is governed by Maxwell equations (ME), namely

$$\nabla \wedge \boldsymbol{E} = -\frac{1}{c}\dot{\boldsymbol{H}}, \tag{2.1}$$

$$\nabla \cdot \boldsymbol{E} = 4\pi\rho, \tag{2.2}$$

$$\nabla \wedge \boldsymbol{H} = \frac{4\pi}{c}\boldsymbol{j} + \frac{1}{c}\dot{\boldsymbol{E}}, \tag{2.3}$$

$$\nabla \cdot \boldsymbol{H} = 0, \tag{2.4}$$

F. Sansò and F. Migliaccio, *Quantum Measurement of Gravity for Geodesists and Geophysicists*, Springer Geophysics,
https://doi.org/10.1007/978-3-030-42838-9_2

where c is the velocity of light in vacuum, partial derivatives with respect to time have been denoted by a dot and ρ, $\boldsymbol{j}$ represent the spatial density of charges and current intensity, respectively, namely

$$\boldsymbol{j}(\boldsymbol{x}, t) = \rho(\boldsymbol{x}, t)\boldsymbol{v}(\boldsymbol{x}, t) . \tag{2.5}$$

Remark 2.1 In classical electrodynamics, charges are never created or destroyed, i.e. they are conserved in time. For any quantity $q(\boldsymbol{x}, t)$, flowing at $\boldsymbol{x}$ with the velocity $\boldsymbol{v}(\boldsymbol{x}, t)$, the conservation equation writes

$$\nabla \cdot (q\boldsymbol{v}) + \dot{q} = 0 . \tag{2.6}$$

In fact, integrating (2.6) over a volume B fixed in time, with boundary S, by applying the Gauss theorem, we get

$$\int_S q(\boldsymbol{x}, t)\boldsymbol{v}(\boldsymbol{x}, t) \cdot \boldsymbol{n}(\boldsymbol{x}) dS + \int_B \frac{\partial q(\boldsymbol{x}, t)}{\partial t} dB \equiv 0 .$$

This equation says that the variation of the quantity q contained in B, namely

$$\frac{d}{dt} \int_B q(\boldsymbol{x}, t) dB \equiv \int_B \frac{\partial q(\boldsymbol{x}, t)}{\partial t} dB ,$$

is the opposite of the total flow of q through the boundary S.

So in the case of charges we must have, recalling (2.5),

$$\nabla \cdot \boldsymbol{j} + \dot{q} = 0 . \tag{2.7}$$

This equation is a consequence of ME; in fact, by taking the divergence of (2.3), taking into account that $\nabla \cdot (\nabla \wedge \boldsymbol{H}) \equiv 0$, and that

$$\nabla \cdot \dot{\boldsymbol{E}} = 4\pi \dot{\rho}$$

from (2.2) we receive

$$\frac{4\pi}{c} \nabla \cdot \boldsymbol{j} + \frac{4\pi}{c} \dot{\rho} = 0 ,$$

so proving (2.6).

Remark 2.2 Physically speaking, there is another source of the magnetic field, namely the magnetic moment $\boldsymbol{\mu}$ of a charged particle. This, however, can only be understood in relation to the spin of the particle, which beyond its name, reminiscent of a classical mechanical quantity, is in reality a quantum variable intrinsic to the particle, that can only be described in the framework of quantum mechanics, as we will see in Sect. 5.6.

ME express the evolution of an e.m. field $(\boldsymbol{E}, \boldsymbol{H})$ in space and time, due to the presence of charges and to their motion. Vice versa, we are also interested in understanding what is the effect of an e.m. field $(\boldsymbol{E}, \boldsymbol{H})$ on a particle, possessing a charge e considered as a point in space.

This effect is expressed in terms of a force $\boldsymbol{F}$ acting on the particle according to Lorentz equation

$$\boldsymbol{F} = e\boldsymbol{E} + \frac{1}{c} e\boldsymbol{v} \wedge \boldsymbol{H} . \tag{2.8}$$

So, if we neglect the effect of the individual particle on $(\boldsymbol{E}, \boldsymbol{H})$, we can say that for a given, external e.m. field, the dynamics of the particle is fixed by (2.8) and Newton's equation, or one of its Lagrangian or Hamiltonian formulations.

2.3 Electromagnetic Waves in Empty Space

Let us assume we have a region Ω of space devoid of charges but seat of an e.m. field, which indeed has been generated elsewhere. Then, in Ω, $(\boldsymbol{E}, \boldsymbol{H})$ satisfy the ME without source, namely

$$\nabla \wedge \boldsymbol{E} = -\frac{1}{c} \dot{\boldsymbol{H}}, \tag{2.9}$$

$$\nabla \cdot \boldsymbol{E} = 0, \tag{2.10}$$

$$\nabla \wedge \boldsymbol{H} = \frac{1}{c}, \dot{\boldsymbol{E}} \tag{2.11}$$

$$\nabla \cdot \boldsymbol{H} = 0. \tag{2.12}$$

From (2.9) we derive

$$\nabla \wedge (\nabla \wedge \boldsymbol{E}) = -\frac{1}{c} \nabla \wedge \dot{\boldsymbol{H}} = -\frac{1}{c} \frac{\partial}{\partial t} (\nabla \wedge \boldsymbol{H}) . \tag{2.13}$$

Let us recall that, for whatever field $\boldsymbol{E}$, the identity holds

$$\nabla \wedge (\nabla \wedge \boldsymbol{E}) = \nabla (\nabla \cdot \boldsymbol{E}) - \Delta \boldsymbol{E} ;$$

in particular for the electrical field $\boldsymbol{E}$ that satisfies (2.10) too, one has

$$\nabla \wedge (\nabla \wedge \boldsymbol{E}) = -\Delta \boldsymbol{E} . \tag{2.14}$$

Moreover, from (2.11) one derives

$$\frac{\partial}{\partial t}\nabla \wedge \boldsymbol{H} = \frac{1}{c}\frac{\partial^2 \boldsymbol{E}}{\partial t^2} . \tag{2.15}$$

Wrapping up in (2.13), one gets

$$\frac{1}{c^2}\frac{\partial^2 \boldsymbol{E}}{\partial t^2} = \Delta \boldsymbol{E} . \tag{2.16}$$

By a complementary reasoning, we see that $\boldsymbol{H}$ satisfies a similar equation, namely

$$\frac{1}{c^2}\frac{\partial^2 \boldsymbol{H}}{\partial t^2} = \Delta \boldsymbol{H} . \tag{2.17}$$

Equations (2.16) and (2.17) are called *wave equations*; the reason of this name can be understood by means of the Example 2.1 that illustrates a family of solutions called *plane e.m. waves*.

Example 2.1 (*Plane waves*) Let us consider the family of functions

$$\boldsymbol{E}(\boldsymbol{x}, t) = \boldsymbol{E}_0 e^{i(\boldsymbol{k}\cdot\boldsymbol{x}-\omega t)} ; \tag{2.18}$$

obviously $\boldsymbol{E}$ is a complex field, $\boldsymbol{E}_0$ a complex constant vector, while $\boldsymbol{k}$ and ω are a real 3D vector and a scalar constant, respectively. Inserting (2.18) into (2.16) one gets, after simplifying,

$$-|\boldsymbol{k}|^2 + \frac{1}{c^2}\omega^2 = 0 . \tag{2.19}$$

This means that, calling $k = |\boldsymbol{k}|$, (2.18) represents a solution of (2.16) (i.e. its real and imaginary parts are both solutions of (2.16)) if the relation

$$\omega = ck \tag{2.20}$$

is satisfied.

Since $e^{i\vartheta}$ is periodical of period 2π, it is convenient to write

$$\begin{cases} \boldsymbol{k} = 2\pi \boldsymbol{p} = 2\pi p \boldsymbol{n} \\ |\boldsymbol{n}| = 1 , \ k = |\boldsymbol{k}| = 2\pi p \end{cases} \tag{2.21}$$

and

$$\omega = 2\pi f = ck = 2\pi c p \tag{2.22}$$

to obtain (2.18) in the form

$$\boldsymbol{E}(\boldsymbol{x}, t) = \boldsymbol{E}_0 e^{i2\pi(p\boldsymbol{n}\cdot\boldsymbol{x} - ft)} . \tag{2.23}$$

It is clear that p has the dimension of the inverse of a length and f of the inverse of a time, so we can put

$$p = \frac{1}{\lambda}\,,\ f = \frac{1}{T}\,, \tag{2.24}$$

and we have

$$\boldsymbol{E}(\boldsymbol{x}, t) = \boldsymbol{E}_0 e^{i2\pi\left(\frac{\boldsymbol{n}\cdot\boldsymbol{x}}{\lambda} - \frac{t}{T}\right)}\,. \tag{2.25}$$

Written in this way, we see immediately that, if we put

$$\boldsymbol{x} = \boldsymbol{x}_0 + \ell\lambda\boldsymbol{n}\ ,\ t = t_0 + mT \tag{2.26}$$

with ℓ, m any integer, we get

$$\boldsymbol{E}(\boldsymbol{x}, t) = \boldsymbol{E}(\boldsymbol{x}_0, t_0)\ ;$$

i.e. function (2.25) is spatially periodic with period λ, along the direction of $\boldsymbol{n}$, and time periodic with period T. Indeed, λ is the wavelength and T the (time) period of the wave (2.25); p and f are called *wavenumber* and *frequency of the wave*, respectively.

Moreover, the whole exponent in (2.18), namely

$$\varphi = \boldsymbol{k}\cdot\boldsymbol{x} - \omega t \tag{2.27}$$

is called the *phase of the wave* at point $\boldsymbol{x}$ and time t.

Finally, by using Eq. (2.22) we have

$$f = \frac{1}{T} = cp = \frac{c}{\lambda}$$

so that we can write (2.23) as

$$\boldsymbol{E}(\boldsymbol{x}, t) = \boldsymbol{E}_0 e^{i2\pi p(\boldsymbol{n}\cdot\boldsymbol{x} - ct)} \tag{2.28}$$

showing that (2.28) is a signal propagating unaltered along the $\boldsymbol{n}$ direction with velocity c; moreover, one has

$$c = \frac{\lambda}{T}\,. \tag{2.29}$$

At this point we can clarify why we have considered the relation (2.20) as the only solution of (2.19) disregarding the other

$$\omega = -ck\ .$$

In fact it is clear that this would correspond just to a wave propagating in the opposite direction $-\boldsymbol{n}$, but with the same velocity, namely we would again find ourselves in the same framework of plane waves theory. However, to complete our analysis, we need to show that (2.18) is a solution of ME, finding also the associated magnetic field $\boldsymbol{H}$.

To do that, we first observe that acting on the function

$$w(\boldsymbol{x}, t) = e^{i(\boldsymbol{k}\cdot\boldsymbol{x}-\omega t)} \tag{2.30}$$

we have the following equivalence of operators:

$$\nabla \sim i\boldsymbol{k}\ , \quad \frac{\partial}{\partial t} \sim -i\omega\ . \tag{2.31}$$

With this simplification ME write

$$\begin{aligned} i\boldsymbol{k} \wedge \boldsymbol{E}_0 w &= -\frac{1}{c}\dot{\boldsymbol{H}} \\ i\boldsymbol{k} \cdot \boldsymbol{E}_0 w &= 0 \\ \nabla \wedge \boldsymbol{H} &= -\frac{i\omega}{c}\boldsymbol{E}_0 w \\ \nabla \cdot \boldsymbol{H} &= 0\ . \end{aligned}$$

It is clear that such equations suggest a solution of the form

$$\boldsymbol{H}(\boldsymbol{x}, \boldsymbol{t}) = \boldsymbol{H}_0 w(\boldsymbol{x}, t)\ , \tag{2.32}$$

so that the above system becomes

$$k\boldsymbol{n} \wedge \boldsymbol{E}_0 = \frac{\omega}{c}\boldsymbol{H}_0, \tag{2.33}$$

$$\boldsymbol{n} \cdot \boldsymbol{E}_0 = 0, \tag{2.34}$$

$$k\boldsymbol{n} \wedge \boldsymbol{H}_0 = -\frac{\omega}{c}\boldsymbol{E}_0, \tag{2.35}$$

$$\boldsymbol{n} \cdot \boldsymbol{H}_0 = 0. \tag{2.36}$$

Recalling that $\omega = ck$ by (2.22), we find that, given any $\boldsymbol{E}_0$ orthogonal to $\boldsymbol{n}$, so that (2.34) is satisfied, $\boldsymbol{H}_0$ is fixed by

$$\boldsymbol{H}_0 = \boldsymbol{n} \wedge \boldsymbol{E}_0\ . \tag{2.37}$$

In fact, such $\boldsymbol{H}_0$ is orthogonal to $\boldsymbol{n}$ too, as well as orthogonal to $\boldsymbol{E}_0$, so that (2.36) is satisfied. Moreover

$$\boldsymbol{k} \wedge (k\boldsymbol{n} \wedge \boldsymbol{H}_0) = -\frac{\omega}{c}\boldsymbol{k} \wedge \boldsymbol{E}_0 = -\frac{\omega^2}{c^2}\boldsymbol{H}_0\ ,$$

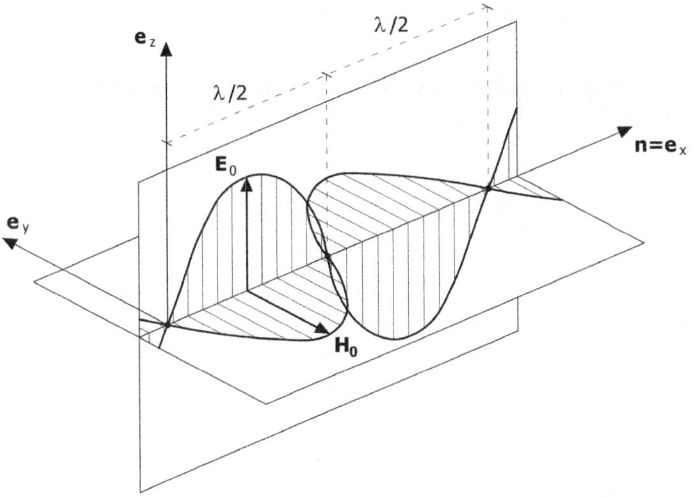

Fig. 2.1 A polarized plan wave

but

$$k\boldsymbol{n} \wedge (k\boldsymbol{n} \wedge \boldsymbol{H}_0) = k^2\boldsymbol{n}(\boldsymbol{n} \cdot \boldsymbol{H}_0) - k^2\boldsymbol{H}_0 \equiv -k^2\boldsymbol{H}_0$$

and, since $k^2 = \frac{\omega^2}{c^2}$, (2.35) is reduced to the identity

$$\boldsymbol{H}_0 \equiv \boldsymbol{H}_0 \, .$$

Summarizing we can say that a plane wave has a geometry fixed by $\boldsymbol{E}_0$ and $\boldsymbol{n}$, which must be orthogonal to one another; then $\boldsymbol{H}_0$ is also fixed by (2.37), namely it is orthogonal to both the propagation direction and the direction of $\boldsymbol{E}_0$ (see Fig. 2.1).

A plane wave with a constant plane in which $\boldsymbol{E}$ vibrates and an orthogonal constant plane in which $\boldsymbol{H}$ vibrates, is said to have a linear polarization. Other polarizations are possible by combining more plane waves, with the same propagation direction. This is a consequence of the linearity of the ME and of the wave equation: the linear combination of two solutions is again a solution of the same equations. So, if we add, for example, a real wave with electric field

$$\boldsymbol{E}_1(\boldsymbol{x}, t) = \boldsymbol{E}_1 \cos(\boldsymbol{x} \cdot \boldsymbol{x} - \omega t)$$

($\boldsymbol{E}_1$ real directed along the x axis)

to a real wave

$$\boldsymbol{E}_2(\boldsymbol{x}, t) = \boldsymbol{E}_2 \sin(\boldsymbol{k} \cdot \boldsymbol{x} - \omega t)$$

($\boldsymbol{E}_2$ real directed along the y axis and such that $|\boldsymbol{E}_2| = |\boldsymbol{E}_1|$), we have indeed a composite real wave

$$\boldsymbol{E}(\boldsymbol{x}, t) = \boldsymbol{E}_1 \cos(\boldsymbol{k} \cdot \boldsymbol{x} - \omega t) + \boldsymbol{E}_2 \sin(\boldsymbol{k} \cdot \boldsymbol{x} - \omega t)$$

such that

$$\begin{aligned} |\boldsymbol{E}|^2 &= |\boldsymbol{E}_1|^2 \cos^2(\boldsymbol{k} \cdot \boldsymbol{x} - \omega t) + |\boldsymbol{E}_2|^2 \sin^2(\boldsymbol{k} \cdot \boldsymbol{x} - \omega t) \\ &= |\boldsymbol{E}_1|^2 = \text{const} ; \end{aligned}$$

so the trajectory of $\boldsymbol{E}(\boldsymbol{x}, t)$ moves in space in the direction of $\boldsymbol{k}$ like a helix on a cylinder, namely one says that the wave is *circularly polarized*. Other polarizations are indeed possible.

2.4 Wave Packets

Plane e.m. waves are not sustainable from the physical point of view. In fact, they fill uniformly the whole empty space, without any source; furthermore a plane wave carries, as we will see, an infinite quantity of energy, which nothing can physically produce. Nevertheless, plane waves are essential in electrodynamics for two reasons: first of all, they provide a very good approximation of certain physical experiments when the empty region where we need to describe the e.m. field is very large with respect to their wavelengths; second, thanks to the superposition principle they can be combined to form new waves with different propagation laws.

Again, the superposition principle is nothing but a direct consequence of the linearity of our equations; so, after a suitable limit process, we can transform a linear combination into an integral to get

$$\boldsymbol{E}(\boldsymbol{x}, t) = \int \widehat{\boldsymbol{E}}(\boldsymbol{k}) e^{i(\boldsymbol{k} \cdot \boldsymbol{x} - \omega t)} d_3 k \, . \tag{2.38}$$

One might ask why not to extend the combination also on different values of the frequency variable ω; the reason is that ω is not a free variable but it is strictly related to k by (2.20), i.e. $\omega = ck$, if we want (2.38) to be a solution of the wave equation (2.16), where c is a fixed constant.

Similarly, we can put

$$\boldsymbol{H}(\boldsymbol{x}, t) = \int \widehat{\boldsymbol{H}}(\boldsymbol{k}) e^{i(\boldsymbol{k} \cdot \boldsymbol{x} - \omega t)} d_3 k \, , \tag{2.39}$$

where again $\omega = ck$ because it has to be a solution of (2.17). Moreover, from (2.9) we derive

$$-\frac{1}{c}\dot{\boldsymbol{H}}(\boldsymbol{x},t) = i\int \frac{\omega}{c}\widehat{\boldsymbol{H}}(\boldsymbol{k})e^{i(\boldsymbol{k}\cdot\boldsymbol{x}-\omega t)}d_3k = \tag{2.40}$$
$$= i\int \boldsymbol{k}\wedge\widehat{\boldsymbol{E}}(\boldsymbol{k})e^{i(\boldsymbol{k}\cdot\boldsymbol{x}-\omega t)}d_3k\ ,$$

namely

$$\widehat{\boldsymbol{H}}(\boldsymbol{k}) = \frac{\boldsymbol{k}}{k}\wedge\widehat{\boldsymbol{E}}(\boldsymbol{k})\ , \tag{2.41}$$

which is strictly analogous to (2.37).

Now let us return to our general expression (2.38) and observe that this is very general, describing a combination of waves of all frequencies travelling in all directions. We want to restrict it to represent a group of waves travelling in the same direction $\boldsymbol{n}$, that we will call a *wave packet*. For this purpose we can always adopt a system of axes with $\boldsymbol{e}_x$ pointing in the direction $\boldsymbol{n}$ and $\boldsymbol{e}_y, \boldsymbol{e}_z$ orthogonal to it. Then, if we specialize (2.38) according to

$$\widehat{\boldsymbol{E}}(\boldsymbol{k}) = \widehat{\boldsymbol{E}}_0(k_x)\delta(k_y)\delta(k_z) \tag{2.42}$$

we see that the two integrals in the y and z directions in (2.38) are cancelled and we are left with

$$\boldsymbol{E}(\boldsymbol{x},t) = \int \widehat{\boldsymbol{E}}(k_x)e^{i(k_x x-\omega t)}dk_x \tag{2.43}$$

or, recalling that in this case $\omega = k_x c$

$$\boldsymbol{E}(\boldsymbol{x},t) = \boldsymbol{E}(x-ct) = \int \widehat{\boldsymbol{E}}(k_x)e^{ik_x(x-ct)}dk_x\ . \tag{2.44}$$

It is easy to verify that (2.44) is a particular solution of the wave equation, which depends on x, t only, i.e. it is constant in y and z, and represents a signal travelling in the positive direction of the x axis with velocity c.

The shape of this signal is obviously governed by the vector function $\widehat{\boldsymbol{E}}(k_x)$, of which we recognize that $\boldsymbol{E}(x-ct)$ is just the Fourier transform.

Example 2.2 We consider here a remarkable example of a wave packet corresponding to the choice

$$\widehat{\boldsymbol{E}}(k_x) = \boldsymbol{E}_0 e^{-\frac{1}{2}\sigma^2 k_x^2}\ . \tag{2.45}$$

We observe that, since $\boldsymbol{E}$ has to be orthogonal to $\boldsymbol{k}$, i.e. to $\boldsymbol{e}_x$ in this case, we must always have

$$\boldsymbol{E}_0\cdot\boldsymbol{e}_x = 0. \tag{2.46}$$

Now we know from the theory of Fourier transform, that (2.44) with (2.45) gives

$$\boldsymbol{E}(x - ct) = \frac{\boldsymbol{E}_0}{\sqrt{2\pi}\sigma} e^{-\frac{1}{2\sigma^2}(x-ct)^2} . \tag{2.47}$$

So a Gaussian in the spectrum $\widehat{\boldsymbol{E}}(\boldsymbol{k}_x)$ is transformed into a Gaussian wave packet in the physical 3D space, travelling along the x- axis. It is useful to remark that if σ is taken as a dispersion index of $\boldsymbol{E}$ along the x-axis, then $\frac{1}{\sigma}$ is clearly the corresponding index of dispersion of the spectrum along the k_x axis.

This fact, which is a general functional property of the Fourier Transform, will be at the basis of the formulation of quantum mechanics principles.

2.5 Electromagnetic Potentials, Lorentz Gauge

We start by proving a fundamental theorem of vector calculus, namely *Helmholtz theorem*.

Theorem 2.1 *Let $\boldsymbol{F}(\boldsymbol{x})$ be any field sufficiently smooth and tending sufficiently fast to zero at infinity with its first derivatives, namely $\left|\frac{\partial}{\partial x_i}\boldsymbol{F}(\boldsymbol{x})\right| = o(|\boldsymbol{x}|^{-2})$, then there are a vector field $\boldsymbol{A}(\boldsymbol{x})$ and a scalar field $\varphi(\boldsymbol{x})$, called vector and scalar potential of $\boldsymbol{F}$ respectively, such that*

$$\boldsymbol{F}(\boldsymbol{x}) = \nabla \wedge \boldsymbol{A} - \nabla\varphi \tag{2.48}$$

Proof The proof is constructive. We start by recalling that

$$\ell_{xy}^{-1} = \frac{1}{|\boldsymbol{x} - \boldsymbol{y}|} \tag{2.49}$$

is a fundamental solution of the Laplace operator, namely

$$\Delta_x \frac{1}{\ell_{xy}} \equiv \Delta_y \frac{1}{\ell_{xy}} = -4\pi\delta(\boldsymbol{x} - \boldsymbol{y}) . \tag{2.50}$$

Accordingly, for any smooth $f(\boldsymbol{x})$ the identity holds

$$\Delta_x \int \frac{f(y)}{\ell_{xy}} dB_y \equiv -4\pi f(x) . \tag{2.51}$$

The same holds as well for any smooth vector field as one verifies componentwise. Next we recall that, thanks to Gauss theorem, the following identity ("integration by parts") holds for any f as above

$$\frac{\partial}{\partial x_i}\int \frac{f(\mathbf{y})}{\ell_{xy}} dB_y = \int \left(-\frac{\partial}{\partial y_i}\frac{1}{\ell_{xy}}\right) f(\mathbf{y}) dB_y = \int \frac{1}{\ell_{xy}}\frac{\partial}{\partial y_i} f(\mathbf{y}) dB_y \ ; \quad (2.52)$$

in fact if $f(\mathbf{y})$ is at least o $(|\mathbf{y}|^{-1})$, the boundary term at infinity implied by Gauss theorem is zero.

By exploiting (2.52) one easily realizes that

$$\nabla_x \int \frac{f(\mathbf{y})}{\ell_{xy}} dB_y = \int \frac{\nabla_y f(\mathbf{y})}{\ell_{xy}} dB_y \quad (2.53)$$

and

$$\nabla_x \wedge \int \frac{\boldsymbol{V}(y)}{\ell_{xy}} dB_y \equiv \int -\nabla_y \left(\frac{1}{\ell_{xy}}\right) \wedge \boldsymbol{V}(y) dB_y \equiv \int \frac{1}{\ell_{xy}} \nabla_y \wedge \boldsymbol{V}(y) dB_y \ . \quad (2.54)$$

We are now ready to prove (2.48), where we put

$$\boldsymbol{A}(\boldsymbol{x}) = \frac{1}{4\pi}\int \frac{\nabla_y \wedge \boldsymbol{F}(\mathbf{y})}{\ell_{xy}} dB_y \quad (2.55)$$

$$\varphi(\boldsymbol{x}) = \frac{1}{4\pi}\int \frac{\nabla_y \cdot \boldsymbol{F}(\mathbf{y})}{\ell_{xy}} dB_y \ . \quad (2.56)$$

In fact, exploiting (2.53) and (2.54), we have

$$\nabla_x \wedge \boldsymbol{A}(\boldsymbol{x}) - \nabla_x \varphi(\boldsymbol{x}) = \frac{1}{4\pi}\int \frac{\nabla_y \wedge (\nabla_y \wedge \boldsymbol{F}(\mathbf{y}))}{\ell_{xy}} dB_y - \frac{1}{4\pi}\int \frac{\nabla_y(\nabla_y \cdot \boldsymbol{F}(\mathbf{y}))}{\ell_{xy}} dB_y =$$

$$= \frac{1}{4\pi}\int \left\{\frac{\nabla_y(\nabla_y \cdot \boldsymbol{F}(\mathbf{y})) - \Delta_y \boldsymbol{F}(\mathbf{y})}{\ell_{xy}} - \frac{\nabla_y(\nabla_y \cdot \boldsymbol{F}(\mathbf{y}))}{\ell_{xy}}\right\} dB_y =$$

$$= -\frac{1}{4\pi}\int \frac{\Delta_y \boldsymbol{F}(y)}{\ell_{xy}} = -\frac{1}{4\pi}\Delta_x \int \frac{\boldsymbol{F}(y)}{\ell_{xy}} dB_y \equiv \boldsymbol{F}(\boldsymbol{x}) \ ;$$

in the last step, (2.51) has been used. □

Corollary 2.1 *The vector potential* $\boldsymbol{A}$ *generating a field* $\boldsymbol{F}$*, is not unique even if we require it to be regular at infinity; the scalar potential* φ *on the contrary is unique if we add the condition that it has to go to zero at infinity.*

Proof If we impose the condition that $(\boldsymbol{A}, \varphi)$ and $(\boldsymbol{A}', \varphi')$ generate the same field $\boldsymbol{F}$, we have

$$\nabla \wedge \boldsymbol{A} - \nabla\varphi \equiv \nabla \wedge \boldsymbol{A}' - \nabla\varphi' \quad (2.57)$$

implying

$$\nabla \wedge (\boldsymbol{A} - \boldsymbol{A}') = \nabla(\varphi - \varphi') . \tag{2.58}$$

Taking the divergence of (2.58) and recalling that, $\forall \boldsymbol{v}$,

$$\nabla \cdot (\nabla \wedge \boldsymbol{v}) \equiv 0 ,$$

we see that

$$0 \equiv \Delta(\varphi - \varphi') ,$$

i.e. $\varphi - \varphi'$ has to be a harmonic function in R^3; since both φ and φ' have to go to zero at infinity, it can only be

$$\varphi \equiv \varphi' . \tag{2.59}$$

But this, with (2.58), implies that

$$\nabla \wedge (\boldsymbol{A} - \boldsymbol{A}') = 0 . \tag{2.60}$$

As it is well-known, a field $\boldsymbol{v}$ that is irrotational, i.e. $\nabla \wedge \boldsymbol{v} = 0$, has a scalar potential namely

$$\boldsymbol{v} = \nabla \psi ,$$

(see also Corollary 1.2).

Therefore, we must have

$$\boldsymbol{A}' = \boldsymbol{A} + \nabla \psi ,$$

where on ψ we impose the only constraint that it is regular and zero at infinity.

So we have proved that the potentials $(\boldsymbol{A}', \varphi')$ and $(\boldsymbol{A}, \varphi)$ generate the same field $\boldsymbol{F}$ if and only if

$$\varphi' = \varphi , \quad \boldsymbol{A}' = \boldsymbol{A} + \nabla \psi \tag{2.61}$$

with ψ arbitrary but going to zero at infinity. One has to remark that by using the (almost) arbitrary function ψ, one can dispose of the term $\nabla \cdot \boldsymbol{A}$; in fact, in the event that $\nabla \cdot \boldsymbol{A} \neq 0$, one can switch to an equivalent $\boldsymbol{A}' = \boldsymbol{A} + \nabla \psi$ and impose

$$\nabla \cdot \boldsymbol{A}' = \nabla \cdot \boldsymbol{A} + \Delta \psi = 0 \tag{2.62}$$

which is indeed possible because (2.62) has the explicit solution

$$\psi = \frac{1}{4\pi}\int \frac{\nabla \cdot \boldsymbol{A}}{\ell_{xy}} dB_y \,. \tag{2.63}$$

Also other constraints are possible; however, one can observe that the explicit form (2.54) of $\boldsymbol{A}$ already satisfies the condition $\nabla \cdot \boldsymbol{A} \equiv 0$, as one can see again exploiting (2.52). □

Corollary 2.2 *A field $\boldsymbol{F}$ is called irrotational if $\nabla \wedge \boldsymbol{F} = 0$ and divergence-free if $\nabla \cdot \boldsymbol{F} = 0$. An irrotational field can always be expressed as*

$$\boldsymbol{F} = -\nabla\varphi \,; \tag{2.64}$$

a divergence-free field can always be expressed as

$$\boldsymbol{F} = \nabla \wedge (\boldsymbol{A} + \nabla\psi) \tag{2.65}$$

with an arbitrary ψ.

Proof Since $\boldsymbol{F}$ can be written as

$$\boldsymbol{F} = \nabla \wedge \boldsymbol{A} - \nabla\varphi \tag{2.66}$$

with a particular $\boldsymbol{A}$ with zero divergence, when $\boldsymbol{F}$ is irrotational by taking the curl of (2.66) we get

$$\nabla \wedge \boldsymbol{F} = \nabla \wedge (\nabla \wedge \boldsymbol{A}) = \nabla(\nabla \cdot \boldsymbol{A}) - \Delta\boldsymbol{A} = -\Delta\boldsymbol{A} \equiv 0 \,;$$

but then it has to be $\boldsymbol{A} = 0$ and (2.64) holds. On the contrary, if $\boldsymbol{F}$ is divergence-free we have

$$\nabla \cdot \boldsymbol{F} = -\Delta\varphi \equiv 0$$

implying that $\varphi = 0$; but then (2.65) holds.

Let us notice that indeed a field $\boldsymbol{F}$ that is at the same time irrotational and divergence-free and zero at infinity has to be identically zero. □

Now, since the argument is general and we will use it in future, we return to the Maxwell Equations with sources and apply the Helmholtz theorem, to obtain the formulation of such equations in terms of e.m. potentials. We start from (2.4) and using Corollary 2.2 we can say that there must be an $\boldsymbol{A}'$ such that

$$\boldsymbol{H} = \nabla \wedge \boldsymbol{A}' \,; \tag{2.67}$$

notice that $\boldsymbol{A}'$ is one among the many potentials that do verify (2.67).

Now from (2.1) and (2.67) we find

$$\nabla \wedge \boldsymbol{E} = -\frac{1}{c}\dot{\boldsymbol{H}} \equiv -\frac{1}{c}\nabla \wedge \dot{\boldsymbol{A}}'$$

i.e.

$$\nabla \wedge \left(\boldsymbol{E} + \frac{1}{c}\dot{\boldsymbol{A}}'\right) = 0 ; \tag{2.68}$$

applying again Corollary 2.2, (2.68) yields

$$\boldsymbol{E} + \frac{1}{c}\dot{\boldsymbol{A}}' = \nabla\varphi'$$

for any scalar field φ'; we write the equation in the form

$$\boldsymbol{E} = -\frac{1}{c}\dot{\boldsymbol{A}}' + \nabla\varphi' . \tag{2.69}$$

Substituting (2.67) and (2.69) into (2.3) and taking into account that

$$\nabla \wedge (\nabla \wedge \boldsymbol{A}') = \nabla(\nabla \cdot \boldsymbol{A}') - \Delta\boldsymbol{A}' ,$$

we derive, after reordering

$$\nabla\left(\nabla \cdot \boldsymbol{A}' - \frac{1}{c}\dot{\varphi}'\right) + \frac{1}{c^2}\ddot{\boldsymbol{A}}' - \Delta\boldsymbol{A}' = \frac{4\pi}{c}j . \tag{2.70}$$

Similarly, substituting (2.69) into (2.3) we get

$$-\frac{1}{c}\nabla \cdot \dot{\boldsymbol{A}}' + \Delta\varphi' = 4\pi\rho ,$$

which we rewrite in the equivalent form

$$-\frac{1}{c}\frac{\partial}{\partial t}\left(\nabla \cdot \boldsymbol{A}' - \frac{1}{c}\dot{\varphi}'\right) - \frac{1}{c^2}\ddot{\varphi}' + \Delta\varphi' = 4\pi\rho . \tag{2.71}$$

Now let us transform our potentials $\boldsymbol{A}'$, φ' into

$$\begin{cases} \boldsymbol{A} = \boldsymbol{A}' + \nabla\psi \\ \varphi = \varphi' + \frac{1}{c}\dot{\psi} \end{cases} \tag{2.72}$$

with ψ an arbitrary function.

It is clear that

$$\begin{cases} \boldsymbol{H} = \nabla \wedge \boldsymbol{A} \\ \boldsymbol{E} = -\frac{1}{c}\dot{\boldsymbol{A}} + \nabla\varphi \end{cases} \tag{2.73}$$

as one verifies by substitution.

So repeating the above reasoning we can claim that $\boldsymbol{A}$, φ do satisfy the two equations

$$\nabla\left(\nabla\cdot\boldsymbol{A} - \frac{1}{c}\dot{\varphi}\right) + \frac{1}{c^2}\ddot{\boldsymbol{A}} - \Delta\boldsymbol{A} = \frac{4\pi}{c}\boldsymbol{j}\,, \tag{2.74}$$

$$-\frac{1}{c}\frac{\partial}{\partial t}\left(\nabla\cdot\boldsymbol{A} - \frac{1}{c}\dot{\varphi}\right) - \frac{1}{c^2}\ddot{\varphi} + \Delta\varphi = 4\pi\rho\,. \tag{2.75}$$

On the other hand, by (2.72)

$$\nabla\dot{\boldsymbol{A}} - \frac{1}{c}\dot{\varphi} = \nabla\dot{\boldsymbol{A}}' - \frac{1}{c}\dot{\varphi}' + \Delta\psi - \frac{1}{c}\ddot{\psi}\,; \tag{2.76}$$

so if the initial choice $\boldsymbol{A}'$, φ' did not satisfy the relation

$$\nabla\dot{\boldsymbol{A}}' - \frac{1}{c}\dot{\varphi}' \equiv 0\,,$$

we can always exploit our arbitrary ψ to impose that $\boldsymbol{A}$, φ do satisfy

$$\nabla\cdot\boldsymbol{A} - \frac{1}{c}\dot{\varphi} \equiv 0\,.$$

Such condition is called the *Lorentz gauge*. Once it is verified, we see from (2.74), (2.75) that $\boldsymbol{A}$, φ do satisfy two wave equations, one with source term $\frac{4\pi}{c}\boldsymbol{j}$, the other with $-4\pi\rho$. In this way we have succeeded in separating the effects of the sources in Maxwell Equations, one for the vector potential only $\boldsymbol{A}$, the other for the scalar potential φ only.

We summarize the above discussion formulating the following theorem.

Theorem 2.2 *Given an e.m. field* $(\boldsymbol{E}, \boldsymbol{H})$ *with sources* $\boldsymbol{j}$ *and* ρ*, there are a vector potential* $\boldsymbol{A}$ *and a scalar potential* φ *such that the following set of equations are satisfied:*

$$\boldsymbol{H} = \nabla\wedge\boldsymbol{A}\,,\quad \frac{1}{c^2}\ddot{\boldsymbol{A}} - \Delta\boldsymbol{A} = \frac{4\pi}{c}\boldsymbol{j}, \tag{2.77}$$

$$\boldsymbol{E} = -\frac{1}{c}\dot{\boldsymbol{A}} + \nabla\varphi\,,\quad \frac{1}{c^2}\ddot{\varphi} - \Delta\varphi = -4\pi\rho. \tag{2.78}$$

$$\nabla \cdot \boldsymbol{A} - \frac{1}{c}\dot{\varphi} \equiv 0 \; (\textit{Lorentz gauge})\,.$$

We remark that, comparing (2.78) with (1.64), the two formulas are the same if we substitute V with $-\varphi$. It follows that the dynamics of a particle in an e.m. field with potentials $\boldsymbol{A}$ and φ is described by the Hamiltonian $\mathscr{H}$ (see (1.62)

$$\mathscr{H} = \frac{1}{2m}|\boldsymbol{p} - \frac{e}{c}\mathbf{A}|^2 - \varphi\,. \tag{2.79}$$

Example 2.3 (*Retarded potentials*) The law governing the evolution of e.m. potentials in terms of wave equations, (2.77), (2.78), can be explicitly solved by formulas resembling the Newton integral, but with a temporal shift; they are called *formulas for retarded potentials* and are

$$\boldsymbol{A}(\boldsymbol{x}, t) = \frac{1}{c}\int_B \frac{\boldsymbol{j}(\mathbf{y}, t-\tau)}{\ell_{xy}} dB_y, \tag{2.80}$$

$$\varphi(\boldsymbol{x}, t) = -\int_B \frac{\rho(\mathbf{y}, t-\tau)}{\ell_{xy}} dB_y, \tag{2.81}$$

where

$$\ell_{xy} = |\mathbf{y} - \boldsymbol{x}|\,,\; \tau(\boldsymbol{x}, \mathbf{y}) = \frac{\ell_{xy}}{c}$$

and B is the volume effectively occupied by charges and currents.

It is enough to prove (2.81); we are going to show that

$$\left(\frac{1}{c^2}\frac{\partial^2}{\partial t^2} - \Delta_x\right)\frac{\rho(\mathbf{y}, t-\tau)}{\ell_{xy}} = 4\pi\rho(\boldsymbol{x}, t)\delta(\boldsymbol{x} - \mathbf{y})\,, \tag{2.82}$$

which, integrated over B, gives (2.78).

We assume that ρ is a smooth function of t and we proceed with the following calculations:

$$\nabla_x \tau = \frac{1}{c}\frac{\boldsymbol{x} - \boldsymbol{y}}{\ell_{xy}} \equiv \frac{1}{c}\boldsymbol{n}\,,\; (|\boldsymbol{n}| = 1)$$

$$\nabla_x \rho(\mathbf{y}, t-\tau) = -\rho_t \nabla_{\boldsymbol{x}}\tau = -\frac{1}{c}\rho_t \boldsymbol{n}$$

$$\Delta_x \rho(\mathbf{y}, t, \tau) = \frac{1}{c^2}\rho_{tt}\boldsymbol{n}\cdot\boldsymbol{n} - \frac{1}{c}\rho_t \nabla_x \cdot \boldsymbol{n} = $$

$$= \frac{1}{c^2}\rho_{tt} - \frac{2}{c}\rho_t \frac{1}{\ell_{xy}}$$

$$\Delta_x \frac{\rho(\mathbf{y}, t-\tau)}{\ell_{xy}} = \frac{(\Delta_x \rho(\mathbf{y}, t-\tau))}{\ell_{xy}} + 2\nabla_x \rho(\mathbf{y}, t-\tau) \cdot \nabla_x \frac{1}{\ell_{xy}} +$$

$$+\rho(\mathbf{y}, t-\tau)\Delta_x \frac{1}{\ell_{xy}} \equiv$$

$$\equiv \frac{1}{c^2} \frac{\rho_{tt}(\mathbf{y}, t-\tau)}{\ell_{xy}} - 4\pi\rho(\mathbf{y}, t-\tau)\delta(\mathbf{x}-\mathbf{y})$$

$$\frac{\partial^2}{\partial t^2} \frac{\rho(\mathbf{y}, t-\tau)}{\ell_{xy}} \equiv \frac{\rho_{tt}(\mathbf{y}, t-\tau)}{\ell_{xy}} .$$

Since the identity

$$\rho(\mathbf{y}, t-\tau)\delta(\mathbf{x}-\mathbf{y}) \equiv \rho(\mathbf{x}, t)\delta(\mathbf{x}-\mathbf{y})$$

holds, we have proved (2.82).

Remark 2.3 In empty space, i.e. when $\rho = 0$, $\boldsymbol{j} = 0$, it is possible to choose another gauge, as we will do in Chap. 6. Namely one can impose

$$\nabla \cdot \boldsymbol{A} = 0 \, , \; \varphi = 0 \tag{2.83}$$

implying

$$\boldsymbol{H} = \nabla \wedge A \, , \; \boldsymbol{E} = -\frac{1}{c}\dot{\boldsymbol{A}} \, , \; \frac{1}{c^2}\ddot{A} - \Delta \boldsymbol{A} = 0 \, . \tag{2.84}$$

That such a choice is consistent can be verified returning directly to (2.9)–(2.12) and showing that $(\boldsymbol{E}, \boldsymbol{H})$ derived from $\boldsymbol{A}$ through (2.84) do satisfy identically those equations.

2.6 Energy Conservation

We consider a system of particles having mass and charge and the e.m. field due to them in an *isolated* part of the space, B. Isolated here means that the volume considered is not subject to force coming from outside, what can be accomplished either by laboratory conditions or by considering a volume with a boundary S so large that no particle nor the e.m. field generated inside, has yet reached it in the time we analyse the system. In such a system each particle, with charge e, will interact with the others through the e.m. field that they generate, according to Lorentz equation (2.8), i.e.

$$\boldsymbol{F}_e = e\boldsymbol{E} + \frac{1}{c} e\boldsymbol{v} \wedge \boldsymbol{H} \, .$$

If we have N particles in the system and

$$n = \frac{dN}{dV} \quad (dV = \text{ volume element})$$

is the particle density, we will have a force density given by

$$\boldsymbol{F} = ne\boldsymbol{E} + \frac{1}{c} ne\boldsymbol{v} \wedge \boldsymbol{H} ,$$

i.e. setting

$$\rho = ne \quad \text{(charge density)}$$

and

$$\boldsymbol{j} = ne\boldsymbol{v} = \rho\boldsymbol{v} \quad \text{(current density)}$$

we find the formula for the force density,

$$\boldsymbol{F} = \rho\boldsymbol{E} + \frac{\boldsymbol{j}}{c} \wedge \boldsymbol{H} . \tag{2.85}$$

If the particles at a point $(\boldsymbol{x})$ move with velocity $\boldsymbol{v}$, we will have that the internal work done by the e.m. field per unit time is

$$\boldsymbol{F} \cdot \boldsymbol{v} = \rho\boldsymbol{E} \cdot \boldsymbol{v} = \boldsymbol{E} \cdot \boldsymbol{j} \tag{2.86}$$

because $\boldsymbol{v} \cdot (\boldsymbol{j} \wedge \boldsymbol{H}) \equiv 0$ as $\boldsymbol{v}$ and $\boldsymbol{j}$ are parallel.

In the system there might be other interactions than the electromagnetic ones and in addition particles will have a kinetic energy. So the total energy of the system will be described by

$$\mathscr{E}_{tot} = \mathscr{E}_{mec} + \mathscr{E}_{em} . \tag{2.87}$$

We admit as a postulate that the total energy $\mathscr{E}_{tot}$ is conserved in time, i.e.

$$\dot{\mathscr{E}}_{tot} = \dot{\mathscr{E}}_{mec} + \dot{\mathscr{E}}_{em} \equiv 0 . \tag{2.88}$$

Since $\boldsymbol{F} \cdot \boldsymbol{v}$ given by (2.86) is the power density of the e.m. interaction, i.e. of the work done by the e.m. field on the system, we will have

$$\dot{\mathscr{E}}_{em} = -\int_B \boldsymbol{E} \cdot \boldsymbol{j} dB . \tag{2.89}$$

Deriving $\boldsymbol{j}$ from (2.3) we write

$$\dot{\mathscr{E}}_{em} = -\frac{c}{4\pi}\int_B \boldsymbol{E}\cdot\nabla\wedge\boldsymbol{H}dB + \frac{1}{4\pi}\int_B \boldsymbol{E}\cdot\dot{\boldsymbol{E}}dB \tag{2.90}$$

Now we consider the identity, descending from (2.1),

$$\frac{c}{4\pi}\boldsymbol{H}\cdot(\nabla\wedge\boldsymbol{E} + \frac{1}{c}\dot{\boldsymbol{H}}) \equiv 0\,. \tag{2.91}$$

Integrating (2.91) over B and adding it to (2.90) yields

$$\dot{\mathscr{E}}_{em} = \frac{c}{4\pi}\int_B (\boldsymbol{H}\cdot\nabla\wedge\boldsymbol{E} - \boldsymbol{E}\cdot\nabla\wedge\boldsymbol{H})dB + \frac{1}{4\pi}\int_B (\boldsymbol{E}\cdot\dot{\boldsymbol{E}} + \boldsymbol{H}\cdot\dot{\boldsymbol{H}})dB\,. \tag{2.92}$$

On the other hand, the following vector identity holds:

$$\nabla\cdot(\boldsymbol{E}\wedge\boldsymbol{H}) = (\nabla\wedge\boldsymbol{E})\cdot\boldsymbol{H} - (\nabla\wedge\boldsymbol{H})\cdot\boldsymbol{E}\,. \tag{2.93}$$

The vector

$$\frac{c}{4\pi}\boldsymbol{E}\wedge\boldsymbol{H} = \boldsymbol{\Pi} \tag{2.94}$$

is called the *Poynting vector*.
So, Eq. (2.92) becomes

$$\begin{aligned}\dot{\mathscr{E}}_{em} &= \int_B \nabla\cdot\boldsymbol{\Pi}dB + \frac{1}{8\pi}\int_B \frac{d}{dt}(|\boldsymbol{E}|^2 + |\boldsymbol{H}|^2)dB = \\ &= \int_S \boldsymbol{\Pi}\cdot\boldsymbol{n}dS + \frac{d}{dt}\int_B \frac{1}{8\pi}(|\boldsymbol{E}|^2 + |\boldsymbol{H}|^2)dB.\end{aligned} \tag{2.95}$$

Since the system is isolated, $\boldsymbol{\Pi}|_S = 0$ and we see that the total energy due to the electromagnetic interaction is spread all over B with a density

$$W = \frac{1}{8\pi}(|\boldsymbol{E}|^2 + |\boldsymbol{H}|^2)\,, \tag{2.96}$$

namely

$$\mathscr{E}_{em} = \int_B WdB\,. \tag{2.97}$$

If the condition $\boldsymbol{\Pi}|_S = 0$ is not satisfied, we obtain from (2.95)

$$\dot{\mathscr{E}}_{em} = \frac{d}{dt}\int_B WdB + \int_B \boldsymbol{\Pi}\cdot\boldsymbol{n}dS \tag{2.98}$$

showing that the flow of the Poynting vector $\boldsymbol{\Pi}$ represents the energy crossing the boundary S, per unit time.

Remark 2.4 Let us take the simple case of a plane wave with a frequency, for instance, in the visible spectrum (wavelengths roughly between 0.4 and 0.7 μm). We know from optics that the intensity of the light I can be defined as the flow of energy per unit area collected by a screen S orthogonal to the direction of propagation. Since from (2.28), (2.37) we have

$$\boldsymbol{E} = \boldsymbol{E}_0 e^{i(\boldsymbol{k}\cdot\boldsymbol{x}-\omega t)} \,, \; \boldsymbol{H} = \boldsymbol{H}_0 e^{i(\boldsymbol{k}\cdot\boldsymbol{x}-\omega t)} \,, \; (\boldsymbol{H}_0 = \boldsymbol{n} \wedge \boldsymbol{E}_0) \,, \tag{2.99}$$

we can say that the energy density

$$W = \frac{1}{8\pi}(|\boldsymbol{E}|^2 + |\boldsymbol{H}|^2) = \frac{1}{8\pi}(E_0^2 + H_0^2) \equiv \frac{1}{4\pi}E_0^2 \tag{2.100}$$

is constant. So, according to (2.96) and (2.97),

$$\begin{aligned} I &= \frac{d\dot{\mathscr{E}}}{dS} = \boldsymbol{\Pi}\cdot\boldsymbol{n} \equiv |\boldsymbol{\Pi}| = \\ &= \frac{c}{4\pi}|\boldsymbol{E}_0 \wedge \boldsymbol{H}_0| = \frac{c}{4\pi}E_0^2 \,. \end{aligned} \tag{2.101}$$

On the other hand, by using (2.100), (2.101) can be written too, as

$$I = cW \,. \tag{2.102}$$

2.7 Momentum Conservation

We now consider the other quantity which we assume to be always conserved in isolated systems, namely the linear momentum. The reasoning will be very similar to the one which we followed to define an energy intrinsic to the e.m. field.

Let us start with the model of an assembly of N particles, each with mass m and charge e; the particle density, as defined in (2.82), is n. Then we know that isolating the behaviour of a single particle i

$$\frac{d}{dt}\boldsymbol{p}_i = \frac{d}{dt}m\,\boldsymbol{v}_i = \boldsymbol{F}_{tot,i} \tag{2.103}$$

where the total forces exerted by the other particles in the system are either of mechanical nature or of e.m. nature, i.e.

$$\boldsymbol{F}_{tot,i} = \boldsymbol{F}_{mec,i} + \boldsymbol{F}_{em,i} \,. \tag{2.104}$$

When the system is isolated, we remarkably have, because of the principle of action-reaction,

$$\sum \boldsymbol{F}_{mec,i} = 0 \,. \tag{2.105}$$

So if we put

$$\boldsymbol{P}_{mec} = \sum m\, \boldsymbol{v}_i = \sum \boldsymbol{p}_i$$

we will have, recalling also (2.85),

$$\begin{aligned}\frac{d}{dt}\boldsymbol{P}_{mec} &= \sum \boldsymbol{F}_{em,i} \cong \int_B n\boldsymbol{F}_{em} dB = \\ &= \int \left(\rho \boldsymbol{E} + \frac{\boldsymbol{j}}{c} \wedge \boldsymbol{H}\right) dB \,. \end{aligned} \tag{2.106}$$

If there were no e.m. interactions, we would have indeed the classical conservation of the mechanical momentum

$$\boldsymbol{P}_{mec} = \text{ const} \,;$$

on the other hand (2.106) shows that $\boldsymbol{P}_{mec}$ is not conserved if there is an e.m. field. But we will show in a moment that

$$\boldsymbol{R}_{em} = \int \left(\rho \boldsymbol{E} + \frac{\boldsymbol{j}}{c} \wedge \boldsymbol{H}\right) dB \,, \tag{2.107}$$

the resultant of e.m. forces, can be written as

$$\boldsymbol{R}_{em} = -\frac{1}{c^2}\frac{d}{dt}\int \boldsymbol{\Pi}\, dB \tag{2.108}$$

with $\boldsymbol{\Pi}$ the Poynting vector, defined in (2.94).

We have then

$$\frac{d}{dt}\left(\boldsymbol{P}_{mec} + \frac{1}{c^2}\int \boldsymbol{\Pi}\, dB\right) = 0 \,, \tag{2.109}$$

showing that, instead of $\boldsymbol{P}$, it is the vector

$$\boldsymbol{P}_{tot} = \boldsymbol{P}_{mec} + \frac{1}{c^2}\int \boldsymbol{\Pi}\, dB \tag{2.110}$$

which is conserved in time. We interpret this fact by stating that to the mechanical momentum $\boldsymbol{P}_{mec}$ we have to add an electromagnetic momentum

$$\boldsymbol{P}_{em} = \frac{1}{c^2}\int \boldsymbol{\Pi} dB = \frac{1}{4\pi c}\int \boldsymbol{E} \wedge \boldsymbol{H} dB \tag{2.111}$$

to obtain a conservation law, that we postulate to hold.

So we have to prove (2.107), (2.108).

From Maxwell Equations we see that

$$4\pi \boldsymbol{R}_{em} = \int \left[\boldsymbol{E}\nabla \cdot \boldsymbol{E} + (\nabla \wedge \boldsymbol{H}) \wedge \boldsymbol{H} - \frac{1}{c}\dot{\boldsymbol{E}} \wedge \boldsymbol{H} \right] dB\,. \tag{2.112}$$

Now we notice that the following identity holds:

$$\int \boldsymbol{H}\nabla \cdot \boldsymbol{H} dB + \int \left(\nabla \wedge \boldsymbol{E} + \frac{1}{c}\dot{\boldsymbol{H}} \right) \wedge \boldsymbol{E} dB \equiv 0 \tag{2.113}$$

because of Eqs. (2.1) and (2.4).

Adding (2.112) and (2.113) we find

$$\begin{aligned} 4\pi \boldsymbol{R}_{em} = & \int (\boldsymbol{E}\nabla \cdot \boldsymbol{E} + \boldsymbol{H}\nabla \cdot \boldsymbol{H}) dB + \\ & + \int [(\nabla \wedge \boldsymbol{H}) \wedge \boldsymbol{H} + (\nabla \wedge \boldsymbol{E}) \wedge \boldsymbol{E})] dB + \\ & + \frac{1}{c}\int (\dot{\boldsymbol{H}} \wedge \boldsymbol{E} - \dot{\boldsymbol{E}} \wedge \boldsymbol{H}) dB\,. \end{aligned} \tag{2.114}$$

By applying the identities

$$(\nabla \wedge \boldsymbol{H}) \wedge \boldsymbol{H} = (\boldsymbol{H} \cdot \nabla)\boldsymbol{H} - \frac{1}{2}\nabla H^2$$

$$(\nabla \wedge \boldsymbol{E}) \wedge \boldsymbol{E} = (\boldsymbol{E} \cdot \nabla)\boldsymbol{E} - \frac{1}{2}\nabla E^2$$

we rewrite (2.114) as

$$\begin{aligned} 4\pi \boldsymbol{R}_{em} = & \int [\boldsymbol{E}(\nabla \cdot \boldsymbol{E}) + (\boldsymbol{E} \cdot \nabla)\boldsymbol{E} + \boldsymbol{H}(\nabla \cdot \boldsymbol{H}) + (\boldsymbol{H} \cdot \nabla)\boldsymbol{H}] dB \\ & - \frac{1}{2}\int \nabla(E^2 + H^2) dB - \frac{1}{c}\frac{d}{dt}\int (\boldsymbol{E} \wedge \boldsymbol{H}) dB\,. \end{aligned} \tag{2.115}$$

We want to show that the first two integrals in (2.115) can be transformed as integrals on the boundary of B, and therefore they are zero for an isolated system.

First notice that, applying the convention of summation on repeated indexes, we have

$$E_i \partial_k E_k + E_k \partial_k E_i = \partial_k (E_k E_i)$$

that we can write in a tensor form

$$\boldsymbol{E}(\nabla \cdot \boldsymbol{E}) + (\boldsymbol{E} \cdot \nabla)\boldsymbol{E} = \nabla \cdot (\boldsymbol{E} \otimes \boldsymbol{E}) . \tag{2.116}$$

Similarly, we have

$$\boldsymbol{H}(\nabla \cdot \boldsymbol{H}) + (\boldsymbol{H} \cdot \nabla)\boldsymbol{H} = \nabla \cdot (\boldsymbol{H} \otimes \boldsymbol{H}) . \tag{2.117}$$

Using (2.116), (2.117), and Gauss' Theorem, we find for the first integral (2.111)

$$\int_B [\boldsymbol{E}(\nabla \cdot \boldsymbol{E}) + (\boldsymbol{E} \cdot \nabla)\boldsymbol{E} + \boldsymbol{H}(\nabla \cdot \boldsymbol{H}) + (\boldsymbol{H} \cdot \nabla)\boldsymbol{H}]dB =$$
$$= \int_B \nabla \cdot [(\boldsymbol{E} \otimes \boldsymbol{E}) + (\boldsymbol{H} \otimes \boldsymbol{H})]dB = \int_S [(\boldsymbol{n} \cdot \boldsymbol{E})\boldsymbol{E} + (\boldsymbol{n} \cdot \boldsymbol{H})\boldsymbol{H}]dS$$

and this has to be zero for an isolated system.

Moreover, by a variant of Gauss' Theorem,

$$\int_B \nabla(E^2 + H^2)dB = \int_S (E^2 + H^2)\boldsymbol{n}dS ,$$

which again has to be zero. So formula (2.108) has been proved and the existence of the e.m. momentum (2.106) has been established.

At the level of notation, let us agree that (2.111) can also be interpreted as describing the total e.m. momentum as a spatial integral of an e.m. momentum density

$$\boldsymbol{P}_{em}(\boldsymbol{x}, t) = \frac{1}{c^2}\boldsymbol{\Pi}(\boldsymbol{x}, t) , \tag{2.118}$$

so that such a momentum can be considered as spread in space, exactly as we did with the energy density (2.96). Therefore, we can summarize the results of the last two sections by stating that every volume element dB occupied by an e.m. field has its own energy WdB and momentum $\boldsymbol{P}_{em}dB$, given by (2.96) and (2.118) respectively.

Remark 2.5 Let us consider a wave packet with a definite propagation direction $\boldsymbol{n}$ and linearly polarized, such that $(\boldsymbol{E}, \boldsymbol{H})$ can be expressed as

$$\begin{cases} \boldsymbol{E} = \boldsymbol{e}_0 \int \widehat{E}(k)e^{ik(x-ct)}dk = \boldsymbol{e}_0 s(x - ct) \\ \boldsymbol{H} = \boldsymbol{n} \wedge \boldsymbol{e}_0 \int \widehat{E}(k)e^{ik(x-ct)}dk = \boldsymbol{n} \wedge \boldsymbol{e}_0 s(x - ct) . \end{cases} \tag{2.119}$$
$$(\boldsymbol{e}_0 \cdot \boldsymbol{n}) = 0$$

Then we immediately notice that indeed

$$E = |\boldsymbol{E}| = |\boldsymbol{H}| = H = |s(x - ct)| ; \tag{2.120}$$

therefore, for the energy density we have

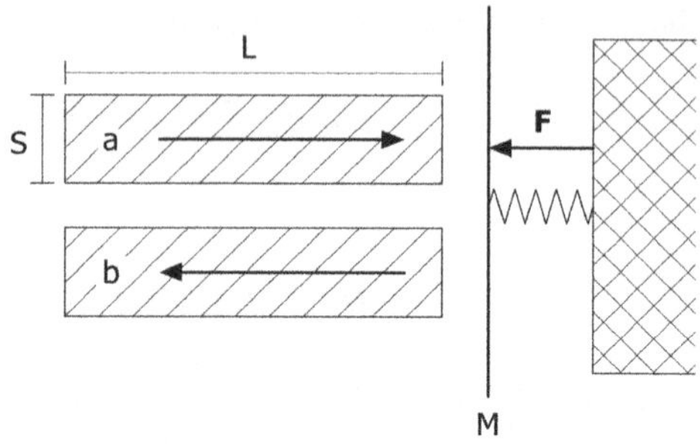

Fig. 2.2 Reflection of a wave packet **a** directed to the mirror M; **b** reflected wave packet; $\boldsymbol{F}$ force required to keep the mirror in equilibrium; the idealized experiment aims at measuring the pressure exerted by the light on M

$$W = \frac{1}{8\pi}(E^2 + H^2) = \frac{1}{4\pi}|s(x - ct)|^2 \,. \tag{2.121}$$

Moreover, the momentum density is given by

$$\boldsymbol{P} = \frac{1}{c^2}\frac{c}{4\pi}\boldsymbol{E} \wedge \boldsymbol{H} = \frac{1}{c}\frac{1}{4\pi}\boldsymbol{n}|s(x - ct)|^2 \tag{2.122}$$

so that the momentum moves along with the propagation direction. We note that the following remarkable relation holds:

$$\boldsymbol{P} = \frac{W}{c}\boldsymbol{n} \,, \tag{2.123}$$

Equation (2.123) slightly generalizes (2.101) that was referred to a monochromatic wave that we will meet again in the chapter on quantum mechanics.

Example 2.4 (*Radiation pressure*) Assume to have a wave packet, like a tube of section S and length L, moving along its directrix (see Fig. 2.2).

The light beam is directed towards a mirror M orthogonal to its velocity and is then reflected. When L is much larger than the wavelength we can consider the beam almost monochromatic and the wave packet can be approximated by a plane wave of a definite frequency. The total momentum of the beam is then given by

$$\boldsymbol{P} = L\,S\boldsymbol{p} = L\,S\frac{W}{c}\boldsymbol{n} \,.$$

When the reflection is completed, after a certain time T, the total momentum is reverted into $-\boldsymbol{P}$, with a momentum variation

$$|\Delta\boldsymbol{P}| = 2LS\frac{W}{c} \,.$$

To a momentum variation a force must correspond impressed by the beam on M, namely

$$|\boldsymbol{F}| = \frac{|\Delta \boldsymbol{P}|}{T} = 2\frac{LSW}{cT} .$$

Since clearly $cT = L$, the above formula becomes

$$|\boldsymbol{F}| = 2SW$$

that corresponds to a pressure on M equal to

$$\text{pressure} = \frac{|\boldsymbol{F}|}{S} = 2W ;$$

this is called the radiation pressure of the beam on M, and it can be measured by the force required to keep M in the same position which is represented as an idealized experiment in Fig. 2.2. We note that, recalling (2.103), the above relation can also be written as

$$\text{pressure} = 2\frac{I}{c} .$$

References

1. Sommerfeld A (1964) Electrodynamics, In: Lecture on Theoretical Physics, Vol 11, Academic Press
2. Feynman RP, Leighton R, Sands M, (2006) The Feinman Lectures on Physics. Definitive Edition. Addison Wesley, Boston
3. Caldirola P, Cirelli R, Prosperi GM (1982) Introduzione alla fisica teorica. UTET, Torino
4. Pauli W (1964) Elettrodinamica. Boringhieri, Torino

Chapter 3
The Crisis of Classical Physics and the Dawn of Quantum Physics

3.1 Introduction

Although at the end of the nineteenth century classical mechanics and the theory of the e.m. field seemed to explain all physical phenomena, it was soon realized that this was not the case because too many experimental data could not be explained by such a theory. Two areas of research and theoretical development were opened at the beginning of the twentieth century. One area studied "high speed" phenomena, that urged the formulation of a new mechanics based on the postulate of the constancy of the velocity of light in the vacuum, which required a deep rethinking of our conception of space and time. This is the relativity theory developed by A. Einstein, which is out of the scope of the present notes. Yet we think it is interesting to underline that it has been natural in this theory to formulate the principle of conservation of energy in a way that clearly showed the equivalence of energy and mass, namely the possibility that a mass could be transformed into radiation, bearing a quantity of energy directly proportional to the mass destroyed.

This direct relation between physical behaviour of masses and radiation, mediated by the form of the Hamiltonian and based on the principle of energy conservation, is also basic for the second development of physics when phenomena at a "small scale" (atomic) were investigated. The "strange" behaviour of high-frequency e.m. radiation, that seemed to behave like a flow of particles, yet displaying a behaviour typical of waves, e.g. interference fringes, pushed the development of a new physics, i.e. quantum physics, due to scientists like E. Schrödinger, W. Heisenberg, P.A.M. Dirac and many others.

In this chapter, we will briefly show examples of physical experiments where the classical theory fails to predict the correct results and that have been used as cornerstones, establishing the principles on which quantum physics has been constructed, as we will see in the next chapter.

Many books report the history of the transition from classical to quantum mechanics, yet for this chapter the book by Caldirola [1] has been used, as well as those by protagonists of this scientific epoch, namely de Broglie [2] and Heisenberg [3].

F. Sansò and F. Migliaccio, *Quantum Measurement of Gravity for Geodesists and Geophysicists*, Springer Geophysics,
https://doi.org/10.1007/978-3-030-42838-9_3

3.2 The Untenable Planetary Model of the Atom

Most of the matter of which we have experience, apart from radioactive material, is clearly in a stable state, stationary in time. Physics has to provide a model for the atoms, that are the constituents of matter, and this model has to justify such a stationarity.

In 1911 Rutherford proposed a planetary model with a positive, massive nucleus of small radius ($10^{-12}/10^{-13}$ cm) around which a number of electrons, with a negative charge balancing that of the nucleus, circulate at a distance of $\sim 10^{-8}$ cm, keeping in equilibrium the attractive electrical force with the centrifugal one.

Notice that the given radius was experimentally estimated on the basis of the scattering of α particles by atoms. We want to show that such a model is untenable in that if we analyse a system with one electron on a circular orbit of radius ε, it creates an e.m. field that produces a positive constant flow of energy on a large sphere with radius R, so that the system can be in a stationary state only if an analogous energy is supplied, to keep the electron in the fixed orbit (Fig. 3.1).

Since this cannot happen for an isolated atom, the model can only predict a collapse of the electrons, namely an instability of the system, contrary to the experience. We will dig out the proof by computing the Poynting vector generated by the electron motion and its flow through the surface of a large sphere showing that such a flow is asymptotically independent of R.

To keep it simple, the computation will be conducted at the first order in ε and $\frac{1}{R}$.

We will use the expression of retarded potentials (2.80), (2.81) taking into account that in this case

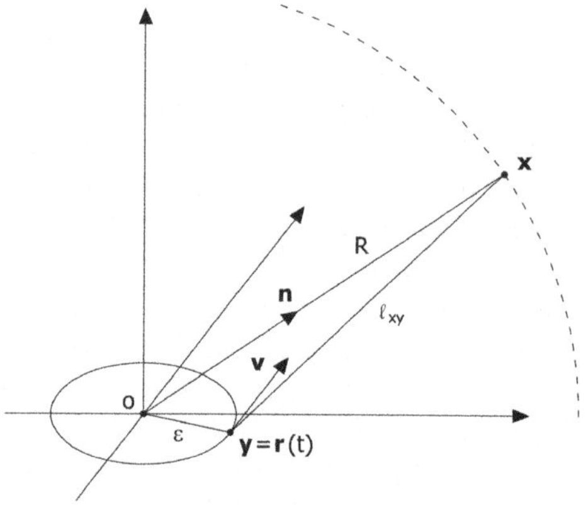

Fig. 3.1 The circular orbit of the electron, with ε = radius of the orbit, R = radius of the sphere where the e.m. field is computed, ℓ_{xy} = distance of the computation point $\boldsymbol{x}$ at time t from the electron at time $t-\tau$

$$\begin{cases} \rho(\boldsymbol{x},t) = e\delta(\boldsymbol{x}-\boldsymbol{y}) \\ \boldsymbol{j}(\boldsymbol{x},t) = e\delta(\boldsymbol{x}-\boldsymbol{y})\boldsymbol{v}(t-\tau), \end{cases} \tag{3.1}$$

$$\boldsymbol{y} = \boldsymbol{r}(t-\tau) \equiv \varepsilon[\cos\omega(t-\tau), \sin\omega(t-\tau), 0]^T, \tag{3.2}$$

$$\tau = \frac{\ell_{xy}}{c} = \frac{|\boldsymbol{x}-\boldsymbol{r}(t-\tau)|}{c}. \tag{3.3}$$

Note must be taken that in this setting,

$$\tau = \tau(\boldsymbol{x},t)$$

is the solution of the implicit Eq. (3.3). However, since we neglect higher order terms, we can put

$$\tau = \frac{R_x}{c}\left[1 + O\left(\varepsilon\frac{1}{R^2}\right)\right] \cong \frac{R_x}{c}. \tag{3.4}$$

Notice that we denote $R = R_x$ because we will need to differentiate with respect to $\boldsymbol{x}$, and therefore R cannot be considered as a constant. With the specifications of (3.1) and (3.2), the retarded potentials are given by

$$\begin{cases} \varphi(\boldsymbol{x},t) = \frac{-e}{\ell_{x,r(t-\tau)}} \\ \boldsymbol{A}(\boldsymbol{x},t) = \frac{e\,\boldsymbol{v}(t-\tau)}{\ell_{x,r(t-\tau)}}. \end{cases} \tag{3.5}$$

Since

$$\varphi(\boldsymbol{x},t) = \frac{-e}{R_x} + O\left(\frac{1}{R^2}\right)$$

$$\nabla_x\varphi(\boldsymbol{x},t) = O\left(\frac{1}{R^2}\right)$$

we see that φ is directly negligible in our computation and we can put (cf. (2.77), (2.78))

$$\begin{cases} \boldsymbol{E} = -\frac{1}{c}\dot{\boldsymbol{A}} \\ \boldsymbol{H} = \nabla \wedge \boldsymbol{A}. \end{cases} \tag{3.6}$$

Now, adopting for $\boldsymbol{A}$ the same approximation as above, we can write

$$\boldsymbol{A} = \frac{e\boldsymbol{v}(t-\tau)}{R_x} + O\left(\frac{1}{R^2}\right) \cong \frac{e\boldsymbol{v}(t-\tau)}{R_x}. \tag{3.7}$$

It is important to understand that $\boldsymbol{A}$ depends on $\boldsymbol{x}$ not only through R_x^{-1}, that if differentiated would produce terms of $O\left(\frac{1}{R^2}\right)$, but also through τ.

So, with the given approximation, we can derive from (3.6)

$$\begin{aligned} \boldsymbol{E} &= -\frac{e}{c}\frac{\boldsymbol{a}(t-\tau)}{R_x}(1-\dot{\tau}) \\ \boldsymbol{H} &= e\frac{\nabla\wedge\boldsymbol{v}(t-\tau)}{R_x} + O\left(\frac{1}{R^2}\right). \\ (\boldsymbol{a}(t-\tau) &= \dot{\boldsymbol{v}}(t-\tau) = -\omega^2\varepsilon\boldsymbol{r}(t-\tau)) \end{aligned}$$

On the other hand, with the approximation (3.4), one has $\dot{\tau} = 0$. Furthermore,

$$\begin{aligned} \nabla\wedge\boldsymbol{v}(t-\tau) &= -\nabla_x\tau\wedge\dot{\boldsymbol{v}}(t-\tau) = \\ &= -\frac{\boldsymbol{n}}{c}\wedge\boldsymbol{a}(t-\tau); \end{aligned}$$

the first equality of the above equation can be verified by computing the curl componentwise, the second by taking into account that

$$\nabla_x R_x = \frac{\boldsymbol{x}}{R_x} = \boldsymbol{n} \quad (|\boldsymbol{n}| = 1). \tag{3.8}$$

Summarizing, one has

$$\boldsymbol{E} = -\frac{e}{c}\frac{\boldsymbol{a}(t-\tau)}{R_x}. \tag{3.9}$$

$$\boldsymbol{H} = -\frac{e}{c}\frac{\boldsymbol{n}\wedge\boldsymbol{a}(t-\tau)}{R_x}. \tag{3.10}$$

and then

$$\begin{aligned} \boldsymbol{\Pi} &= \frac{4\pi}{c}\boldsymbol{E}\wedge\boldsymbol{H} = \frac{e^2}{4\pi c}\frac{\boldsymbol{a}(t-\tau)\wedge[\boldsymbol{n}\wedge\boldsymbol{a}(t-\tau)]}{R_x^2} = \\ &= \frac{e^2}{4\pi c}\frac{|\boldsymbol{a}(t-\tau)|^2\boldsymbol{n} - [\boldsymbol{a}(t-\tau)\cdot\boldsymbol{n}]\ \boldsymbol{a}(t-\tau)}{R_x^2}. \end{aligned}$$

Therefore, since $|\boldsymbol{a}|^2 = \omega^2\varepsilon$ and calling ϑ the angle between $\boldsymbol{a}$ and $\boldsymbol{n}$,

$$\boldsymbol{\Pi}\cdot\boldsymbol{n} = \frac{e^2}{4\pi c}\omega^2\varepsilon\frac{1-\cos^2\vartheta}{R_x^2}. \tag{3.11}$$

The flow of $\boldsymbol{\Pi}$ through the sphere is easily computed if we use a coordinate system with the pole aligned with $\boldsymbol{a}$, so that

$$\int \boldsymbol{\Pi} \cdot \boldsymbol{n} dS = \frac{e^2}{4\pi c} \frac{\omega^2 \varepsilon}{R_x^2} \int_0^{2\pi} d\lambda \int_0^{\Pi} (1 - \cos^2 \vartheta) R_x^2 \sin \vartheta d\vartheta =$$

$$= \frac{e^2 \omega^2 \varepsilon}{6c}. \tag{3.12}$$

As we can see, also recalling (2.96), the flow of energy per unit time is constant, as we wanted to prove.

3.3 The Birth of Quanta

In 1900 Max Planck, studying the radiation of a so-called black body, gave birth to the idea of quanta of Action. Let us remember that, mechanically speaking, the action is the integral of energy in time, or the line integral of the momentum in space.

It has been proved that such a radiation is the same as the one which is found within a metallic cavity when it reaches the equilibrium, namely the radiation absorbed by the electrons in a time dt is the same as that re-emitted in the same time by the walls into the cavity.

Some reasoning of thermodynamic nature and a further computation based on classical mechanics (for the electrons) and classical electrodynamics, has led Rayleigh to formulate a law for the density of radiation energy, for a frequency interval $d\nu$ in a cavity maintained at (absolute) temperature T, that writes

$$e(\nu, T) d\nu = \frac{8\pi k}{c^3} \nu^2 T d\nu \tag{3.13}$$

where k is the Boltzmann constant, with value 1.37×10^{-10} in cgs units (T in Kelvin).

Although (3.13) respects some requirements imposed by thermodynamics, for instance, it is true that $e(\nu, T)$ has the shape $\nu^3 F\left(\frac{\nu}{T}\right)$ for some function F, it is obviously unacceptable for two reasons. The first is that indeed the total energy derived by integrating in $d\nu$ (3.13) over all frequencies on the interval $[0, +\infty)$, would be infinite. But even more, because one can study experimentally the energy spectrum, i.e. the function $e(\nu, T)$ as function of ν for a constant T, obtaining the plot of Fig. 3.2 (solid line), where the law (3.13) is also represented by a dashed line.

As we see, apart from a range of small frequencies the physical reality dramatically departs from formula (3.13) derived by applying the classical physics laws.

Max Planck, in an effort to explain the empirical law of Fig. 3.2, showed that it can be obtained only by making the revolutionary hypothesis that the exchange of energy between electrons and e.m. radiation should happen by multiples of a "minimum energy grain" ε that leads to the formula

$$e(\nu, T) = \frac{8\pi}{c^3} \nu^2 \frac{\varepsilon}{e^{\frac{\varepsilon}{kT}} - 1}. \tag{3.14}$$

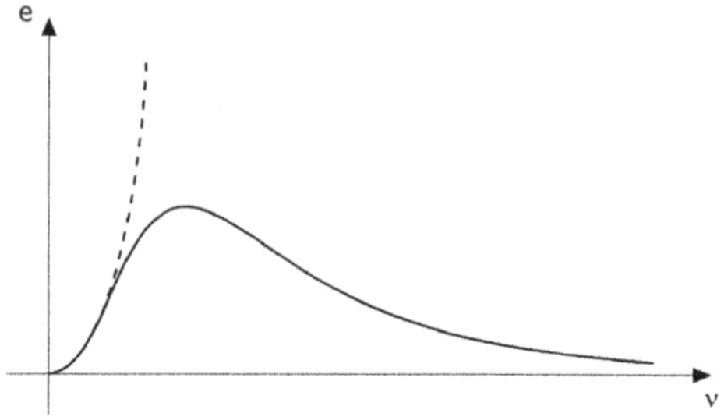

Fig. 3.2 The experimental energy spectrum of a black body (solid line) and the Rayleigh law (dashed line)

On the other hand, considering that it has to be $e(\nu, T) = \nu^3 F\left(\frac{\nu}{T}\right)$, in (3.14) we are forced to put

$$\varepsilon = h\nu \tag{3.15}$$

with h a universal constant, named after the author as the *Planck constant*. Its numerical value, determined from the shape of the empirical spectrum, in the cgs system is

$$h = 6.6 \times 10^{-27} \text{erg s}.$$

It has to be noticed that for small values of the frequency, we recover the expression (3.13) that was known to be correct in that range.

On the other hand, the idea that the exchange of energy between radiation and matter should be explained by some discrete, i.e. not continuous, mechanism was very well present in experimental physics in those years, as energy spectra of radiation emitted by heated matter showed very marked lines at frequencies that were depending only on the kind of matter used in the experiment.

The success of the theory of quanta was then at the same time requiring an investigation on why and how the energy could be concentrated in packets in the e.m. radiation and why and how the structure of the atom seemed to allow an energy exchange only at well-defined frequencies.

3.4 The Electromagnetic Radiation Behaves Like a Flow of Particles

In the eighteenth century I. Newton made the hypothesis that light (a particularly small section of the e.m. spectrum) would be a flow of particles, subject to the law of mechanics. However, in the same century C. Huygens elaborated the theory that light was an undulatory field explaining, for instance, the phenomenon of simultaneous reflection and refraction of a beam. After one century of debates T. Young, and a little later A. Fresnel showed that only the undulatory hypothesis could explain interference and diffraction. The explanation was so strong that a kind of equivalence, interference ≡ waves, has been firmly stated. The theory of J.C. Maxwell, shortly recalled in Chap. 2, gave a unified view of all the e.m. phenomena, ranging from the frequency of radio waves, through the visible spectrum, up to the frequencies of X and γ rays. So it was astonishing for physicists to discover that the empirical evidence of quanta could revive the corpuscular theory.

This was put forward by A. Einstein in 1905, who maintained that the light would be constituted by localized packets of energy, namely light particles that he called photons. Such particles, according to Einstein's hypothesis, bear a discrete quantity of energy according to Planck law

$$E = h\nu. \tag{3.16}$$

There is a lot of empirical evidence of the correctness of the hypothesis that photons exist; here we report only a few of the most famous experiments that at the beginning of the twentieth century were crucial for its proof.

The photoelectric effect. It was already known since 1887, thanks to the work of H.R. Hertz, that a metallic plate irradiated with light would emit a flow of electrons. The kinetic energy T of such electrons has been studied and found independent of the intensity of the light, but rather following a linear law as function of the frequency

$$T \equiv C(\nu - \nu_0) \quad (\nu > \nu_o). \tag{3.17}$$

The frequency ν_0 gives rise to a sill of energy $E_0 = C\nu_0$, corresponding to the minimum amount of work necessary to extract photons form the metal. Such a ν_0 depends only on the metal in question. So the energy of the incident wave, i.e. its intensity, has no effect on the kinetic energy of the expelled electrons, but only on their number.

The constant C can be experimentally determined. In 1916 R. Millikan found that $C = h$ the Planck constant. This proves that the absorption of energy from the light is done by quanta.

The upper limit of the X-ray spectrum. An X-ray tube works by producing electrons from a cathode, letting them hit an anode positively charged, falling through an electric potential difference V so that electrons acquire a kinetic energy equal to eV. The electrons then interact with the matter producing an X-ray e.m. flow. The spectrum of the emitted X rays has two components. One is discrete in the form of intense lines, corresponding to an interaction with electrons bound to the atoms; this will be discussed in the next section. The other component, less intense, is continuous, corresponding to electrons free to run in the metal. This component, however, has a sharp maximum ν_m following the law of Duane and Hunt, namely

$$\nu_m = \frac{eV}{h}.$$

This is interpreted by saying that the maximum energy captured by the X-rays from the falling electrons is eV, which is then emitted in terms of quanta of energy $h\nu_m = eV$. This proves that the emission of e.m. radiation is done by quanta too.

The Compton effect. In Sect. 2.7 we have shown that not only an e.m. wave carries an energy density, proportional to the intensity of the field, but it also carries a momentum density satisfying Eq. (see (2.117))

$$|\boldsymbol{p}| = \frac{1}{4\pi c}|\boldsymbol{E} \wedge \boldsymbol{H}| = \frac{W}{c}.$$

Therefore, when we formulate the hypothesis that light is constituted by "particles", i.e. photons, we expect them to show some behaviour typical of a "localized" momentum transfer, i.e. different from the radiation pressure presented in Example 2.4, which would be compatible with an undulatory concept too. A typical mechanical interaction between two particles could be an elastic collision, which is determined by the conservation of energy and momentum. This is what was experimentally found by A.H. Compton in 1923. The idea is shortly illustrated in Fig. 3.3. A beam of X-rays, namely a radiation having more energetic quanta than that of visible light, is passed through a Wilson chamber, i.e. a chamber where water vapour is kept close to the saturation point. It is well-known that any charged particle passing through the chamber leaves a fairly visible track of fog along its trajectory. However, this does not happen with an e.m. wave. Nevertheless, when a photon hits an electron we can see its trail departing from the direction of the incoming wave, with an angle φ (see Fig. 3.3). Indeed we cannot directly see the bouncing photon, yet when the phenomenon takes place many times, it can happen that the photon in the new direction collides once again with another electron, that then becomes visible. In this case we can also see what is the angle ϑ, under which the photon has bounced (see Fig. 3.3).

Writing the equations of conservation of momentum and energy and taking into account that for the photon it is

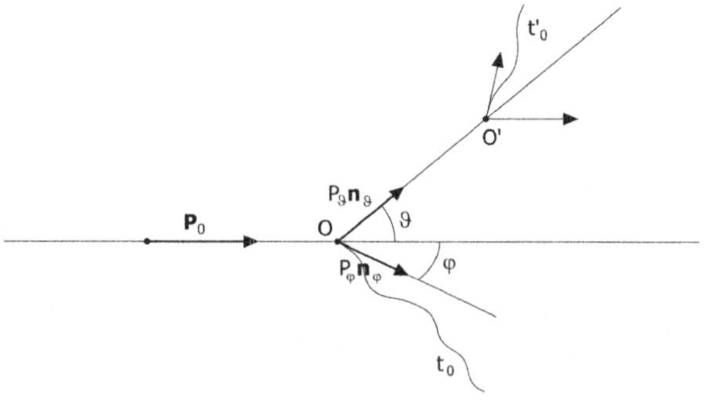

Fig. 3.3 Collision of a photon and an electron at point O; $|\boldsymbol{p}_0| = \frac{E_0}{c}$ moment of the incoming photon; p_ϑ moment of the photon bouncing under the angle ϑ; $p_\varphi = mv$ moment of the visible electron after the primary collision with the photon at O; O' position of the secondary collision, shown by the secondary electron track $t_{0'}$

$$E_0 = h\nu_0 \; ; \; p_0 = \frac{h\nu_0}{c}$$
$$E_\vartheta = h\nu_\vartheta \; ; \; p_\vartheta = \frac{h\nu_\vartheta}{c}$$

while for the electron it is

$$E_\varphi = \frac{1}{2}mv^2 \; ; \; p_\varphi = mv,$$

one obtains

$$\frac{h\nu_0}{c} = \frac{h\nu_\vartheta}{c}\cos\vartheta + mv\cos\varphi$$
$$0 = \frac{h\nu_\vartheta}{c}\sin\vartheta - mv\sin\varphi$$
$$E_0 = h\nu_0 = E_\vartheta + E_\varphi = h\nu_\vartheta + \frac{1}{2}mv^2.$$

Knowing ν_0 (the incoming frequency), m (the mass of the electron), one can derive ν_ϑ and v from the first two equations and put them into the third. In this way one obtains a relation between ϑ and φ. Since those two angles can be measured by taking pictures of the Wilson chamber, one can show that such a relation is well satisfied by data, thus confirming the validity of the theory of the Compton effect.

3.5 The Quantized Atom of Niels Bohr

In studying the interaction between e.m. field and matter, a particular mention has to be done of spectroscopy experiments. The matter, in particular a gas, crossed by waves of different frequencies, absorbs energy according to the quantum law $E = h\nu$ only in correspondence of well-defined frequencies that are seen as lines in the spectrum of the transmitted wave. Reciprocally a heated gas, so that its molecules increase their kinetic energy, starts emitting a radiation but only at well-defined frequencies. Emission and absorption happen at the same lines of the spectrum, namely at the same energies/frequencies which are characteristic of the medium.

Let us discuss the simplest case, namely that of hydrogen with atoms composed only by one proton in the nucleus and one electron. We will reconstruct, in elementary terms, the reasoning done by N. Bohr in 1913.

The idea was to treat the system as based on classical principles, but adding the experimentally ascertained fact that energy could be emitted/absorbed only by quanta $h\nu_{nm}$ that were considered to make the system jump from a stationary state E_n to another stationary state E_m, such that

$$h\nu_{nm} = E_m - E_n. \tag{3.18}$$

The target was to find an explanation of Balmer's empirical law, claiming that for hydrogen

$$\nu_{nm} = R\left(\frac{1}{n^2} - \frac{1}{m^2}\right). \tag{3.19}$$

If we combine (3.18) and (3.19) we see that we have to explain why the stable levels of energy have the form

$$E_n = -\frac{hR}{n^2}. \tag{3.20}$$

To simplify notation, in the subsequent computation we will denote by c any constant value; so c will represent in reality different constants.

We start by stating that the atom behaves as if the electron e follows a circular orbit of radius r. By equating centrifugal and centripetal forces we have

$$m\omega^2 r = \frac{e^2}{r^2} \quad \left(\omega = \frac{v}{r}\right) \tag{3.21}$$

from which the well-known Kepler law follows

$$r = cT^{2/3}, \tag{3.22}$$

with T the orbital period.

On the other hand the classical energy of the electron would then be, with the help of (3.21),

$$E = \frac{1}{2}mv^2 - \frac{e^2}{r} = \frac{1}{2}m\omega^2 r^2 - \frac{e^2}{r} = -\frac{1}{2}\frac{e^2}{r}. \tag{3.23}$$

By using (3.22) and (3.23) we get

$$-E = cT^{-2/3}. \tag{3.24}$$

Now T is the period of the orbit, i.e. $\nu = T^{-1}$ is the orbital frequency. If an e.m. quantum is absorbed/emitted by the electron, this must have the same frequency, i.e. changes in energy are not continuous but can happen only in multiples of the quantum $h\nu$.

Then we can hypothesize that

$$-E_n = nh\nu \tag{3.25}$$

or

$$-E_nT = nh\nu T = nh. \tag{3.26}$$

Finally, (3.26) combined with (3.24) gives

$$cT^{1/3} = nh \Rightarrow E = -\frac{c}{n^2}, \tag{3.27}$$

as we wanted to prove.

We will return to this example, on the basis of a much more structured theory in the next chapter.

3.6 Overcoming the Dualism Waves–Particles. The "Waves of Matter" of Louis de Broglie

We have seen in Sect. 3.4 that e.m. waves can behave like a flow of particles (photons) giving a number of examples where the behaviour was clearly highlighted in experiments. On the other hand, we know that light can display phenomena of diffraction and interference that were taken as "proof" that light has an undulatory nature. A significant contribution to the discussion on this alternative has been given by Louis de Broglie, who in fact inverted the argument predicting, years before an empirical proof, that vice versa also matter could display an undulatory behaviour.

But let us start with the nature of the light. The solution of the dichotomy was that wave and particle are two faces of the same physical object and the connection between the two is that the intensity of the wave at a certain place (i.e. the square of the modulus of the field, representing the energy density, see (2.96)) is proportional to the probability density of the particle in that place. This concept becomes much clearer if one thinks of a light beam with very small intensity, such that the mean volume occupied by a single photon is macroscopic; is this enough to give to the beam a particle-like behaviour? The answer is no. Experience shows that the "trajectory" of the particle is one of the many that it could follow, according to the probability evolution that is governed by the wave dynamics, but we cannot know a priori which of those will be the real one.

This proves that the undulatory dynamics of the wave is part of the physics of light and not just an envelop to describe the statistics of a cloud of particles. An example describing an (idealized) experiment by Young will help to understand the concept.

Example 3.1 Consider a screen Σ with two thin slits S_2, S_2 lighted from behind with a uniform beam of light (see Fig. 3.4); a photographic plate Π is put parallel to Σ at a distance D.

The particle-like picture. If we take the beam as a flow of particles, with a uniform distribution, but following mechanical rules, we would expect that only the particles falling through S_1 and S_2 would then hit Π, leaving an image with two maxima only centred at points I_1, I_2 which are the orthogonal projection of S_1, S_2 on Π. If the

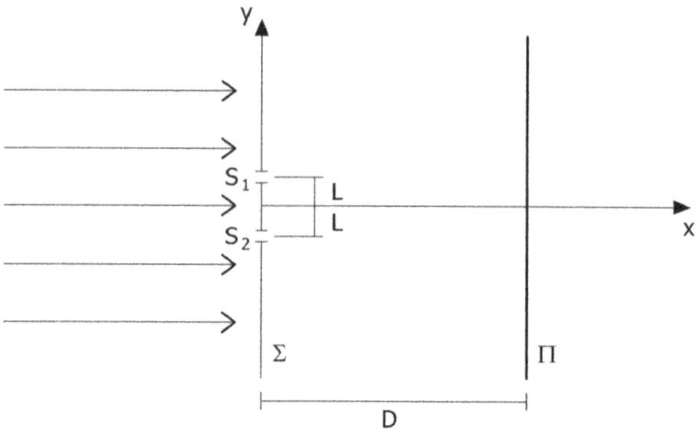

Fig. 3.4 The set up of Young's experiment. In $y < 0$ there is a plane e.m. wave incoming (light); the two slits S_1, S_2 on the screen Σ are separated by a distance $2L$; Π is the photographic plate recording the light on the semispace $y > 0$

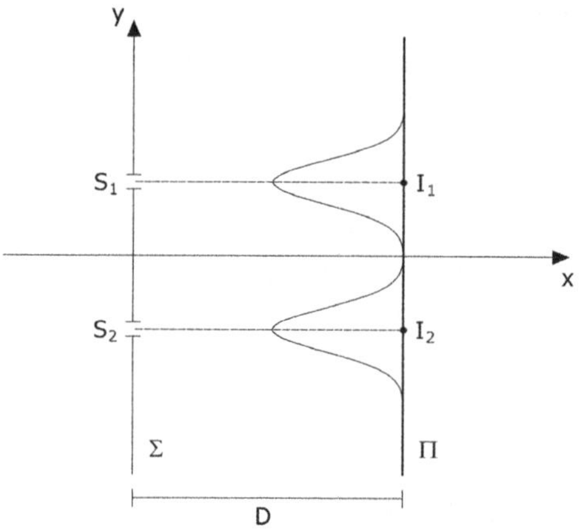

Fig. 3.5 The light distribution on Π if the non-interacting particle dynamics were true

intensity of the flow is decreased, such that only one photon could fall on Σ after a measurable time interval, indeed we would register arrivals on Π well separated in time, but that cumulating would reproduce an image corresponding to the two maxima in Fig. 3.5.

Note that the separation between photons in this case supports the hypothesis that the particles are non-interacting.

The wave-like picture (interference). Now let us consider the same experiment described in terms of propagation of the e.m. field. The result is displayed in Fig. 3.6.

The two slits S_1, S_2 act as new sources of waves, perfectly in phase because Σ is hit by a plane wave orthogonal to it. So the two rays r_1 and r_2 that converge in Q have a phase difference given by

$$\Delta\varphi = \frac{2\pi}{\lambda}[r_1 - r_2], \tag{3.28}$$

with λ the wavelength of the beam illuminating Σ.

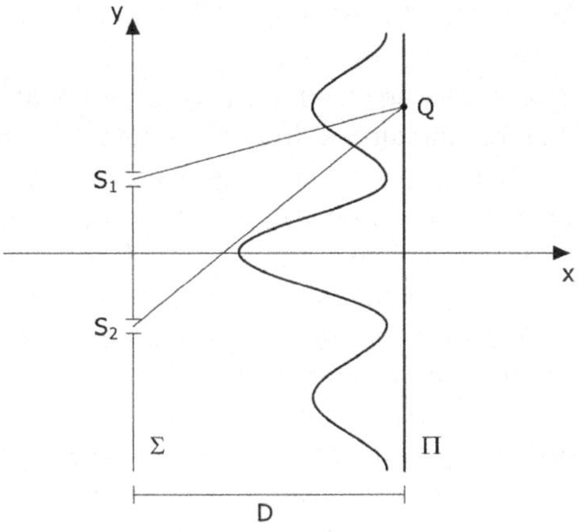

Fig. 3.6 Illustration of the interferences fringes according to the undulatory hypothesis

On the other hand, if y and L are relatively smaller than D, one has approximately

$$r_1 - r_2 = \sqrt{D^2 + (y-L)^2} - \sqrt{D^2 + (y+L)^2} \cong$$
$$\cong \left[D + \frac{1}{2}\frac{(y-L)^2}{D}\right] - \left[D + \frac{1}{2}\frac{(y+L)^2}{D}\right] = -\frac{2Ly}{D}.$$

So it is

$$\Delta\varphi \cong -\frac{4\pi Ly}{\lambda D}. \tag{3.29}$$

When $\Delta\varphi = 2n\pi$ the two waves travelling along r_1, r_2 arrive in phase and the propagated fields add to one another producing a maximum of the intensity; when $\Delta\varphi = (2n+1)\pi$ the phases of the two rays are in opposition and we have a minimum of the intensity. Note that, since from (3.29) the maximum is at $\Delta\varphi = 0$ corresponding to $y = 0$ and the first minimum is at $y = \frac{D\lambda}{4L}$, by measuring the distance between bright and dim stripes we have the possibility of deriving the wavelength λ, if this is not known.

Again, by sensibly reducing the intensity of the beam we would see photons falling on Π one at a time, but cumulating to reproduce the image of Fig. 3.6. Yet the individual photon falls on points Q according to a probability law that is the image on Π of Fig. 3.6, duly normalized. This means that the dynamics of the particle has to be described in terms of wave propagation and probability.

It was exactly this consideration, together with the deep analogy between the principle of least action in mechanics with the Fermat principle governing linear optics, that led Louis de Broglie to postulate that for particles too, there could be an associated material wave, which he called $\psi(\boldsymbol{x}, t)$, following an undulatory dynamics so that also particles should display some wave-like behaviour, like the Young interference presented above.

For e.m. fields interference experiments were carried out with crystals that, with their regular atomic structure, could substitute the Young slits. For instance, in 1912 von Laue proved that X-rays were producing interference in this way. So the experimental verification of de Broglie's hypothesis was carried out with a similar apparatus by Devisson and Germer in 1927, by using a beam of electrons, brought all at the same level of kinetic energy. This is in fact, in terms of particles, the analogous of a plane wave, or better we expect that the material wave ψ of a monokinetic electron beam would be a plane wave. The experiment proved the existence of interference of the electrons, as predicted by de Broglie. But even more, since by measuring the distances of the fringes it is possible to derive the wavelength λ, it was of great importance to compare it with the theoretical value derived from the hypothesis of undulatory mechanics.

We could ask: what is the wavelength associated with an electron travelling with velocity v, i.e. momentum $p = mv$ and energy $E = \frac{1}{2}mv^2$? The answer comes from the mentioned analogy between undulatory mechanics and optics.

Let us recall that for an e.m. wave one has

$$p = \frac{E}{c}$$

that, for a quantum of energy $E = h\nu$, can be written as

$$p = \frac{h\nu}{c} = \frac{h}{\lambda}; \tag{3.30}$$

then generalizing to electrons one can assume that the associated wavelength is given by

$$\lambda = \frac{h}{p} = \frac{h}{\sqrt{2mE}}. \tag{3.31}$$

Since the energy E of the electron is generated by accelerating it through an electrostatic potential, i.e. $E = eV$, it is known and (3.31) can be computed, when we know mass and charge of the electron too.

The result was confirmed by the experiments and this is one of the important empirical foundations of the building of quantum mechanics.

3.7 The Heisenberg Indetermination Principle

We report here two of the many "ideal" experiments that led W. Heisenberg to formulate a principle that is a cornerstone on which quantum mechanics has been formally constructed.

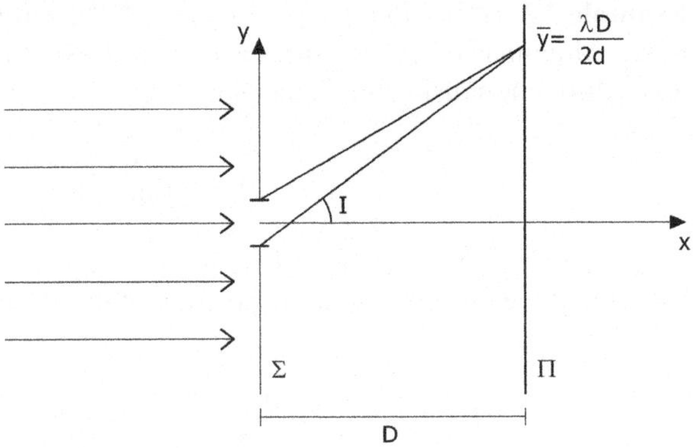

Fig. 3.7 Diffraction by the slit of aperture d, the maximum illumination is at $y = 0$, the first dim fringe at $y = \frac{\lambda D}{2d}$

Example 3.2 *(Photon diffraction)* Classical diffraction of light by a slit is a phenomenon very similar to interference, as described in Example 3.1. The main difference here is that there is only one slit, and its borders act as pointwise sources of new rays, i.e. we have $2L = d$ (see Fig. 3.7). Therefore, we can make use of the simple computation leading to (3.29) to conclude that the distance between the central lighted line and the first dim fringe is $\overline{y} = \frac{\lambda D}{2d}$. Accordingly, the rays that have an inclination

$$tgI < \frac{\overline{y}}{D} = \frac{\lambda}{2d} \tag{3.32}$$

are those falling on the highest probability area. If we think of rays in terms of probable trajectories of photons, possessing a momentum $p = \frac{h}{\lambda}$, we can roughly say that the y component of the momentum is

$$p_y < \frac{h}{\lambda} \cdot \frac{\lambda}{2d} = \frac{h}{2d} = p_{\overline{y}}. \tag{3.33}$$

Formula (3.33) gives the order of magnitude of $p_{\overline{y}}$ immediately after crossing the slit, as before p_y was equal to zero. So (3.33) gives an idea of the uncertainty in p_y of photons,

$$\max p_y = \Delta p_y = p_{\overline{y}} = \frac{h}{2d}, \tag{3.34}$$

after we have determined their position by means of the slit, i.e. with an uncertainty $\Delta y \sim d$. But then we have immediately, after the measurement of the position,

$$\Delta y \Delta p_y \cong \frac{1}{2} h. \tag{3.35}$$

Example 3.3 *(Heisenberg's microscope)* An electron initially at rest on the x-axis is back-illuminated by light with wavelength λ (See Fig. 3.8). Its image on the screen S is a diffraction disk with diameter δ

$$\delta = \frac{\lambda}{2\sin\varepsilon}; \tag{3.36}$$

ε is the angle under which the lens is seen from P. So we can assume that δ is a measure of the uncertainty of the position of the electron in x, i.e.

$$\Delta x = \delta = \frac{\lambda}{2\sin\varepsilon}. \tag{3.37}$$

At the same time, a photon hitting the electron will be generally deviated, while the electron itself will acquire a certain momentum in the (x, y) plane with component (p_{ex}, p_{ey}), related to the momentum $p_0 = \frac{h\nu}{c}$ of the incoming light by the equations of Compton's effect. Notice that the scattered photon is collected by the lens L only if the scattering angle ϑ is smaller than ε. One can prove that the momentum gained by the electron is small compared to that of the photon, i.e. the momentum $\frac{h\nu'}{c}$ of the photon after the impact is almost equal to $\frac{h\nu}{c}$. In this case the conservation of the momentum in the x direction tells us that

$$p_x = p_e \sin\alpha = \frac{h\nu'}{c}\sin\vartheta \le \frac{h\nu}{c}\sin\varepsilon. \tag{3.38}$$

Since p_x was zero in the initial state, we have that max $p_x = \Delta p_x$ is given by

$$\Delta p_x \cong \frac{h\nu}{c}\sin\varepsilon. \tag{3.39}$$

Combining (3.37) and (3.39) one gets

$$\Delta x \Delta p_x \cong \frac{\lambda}{2\sin\varepsilon}\frac{h\nu}{c}\sin\varepsilon = \frac{1}{2}h. \tag{3.40}$$

Other examples could be provided, all leading to the same conclusion that the accuracy in determining the position of a particle cannot be increased unless we simultaneously decrease our knowledge of the momentum of the particle and vice versa. The same happens if we try to determine simultaneously the energy of the state of a particle and the time when such a state was achieved.

All that has driven W. Heisenberg to formulate a general principle concerning the physical state of a system and the possibility of determining its dynamical variables.

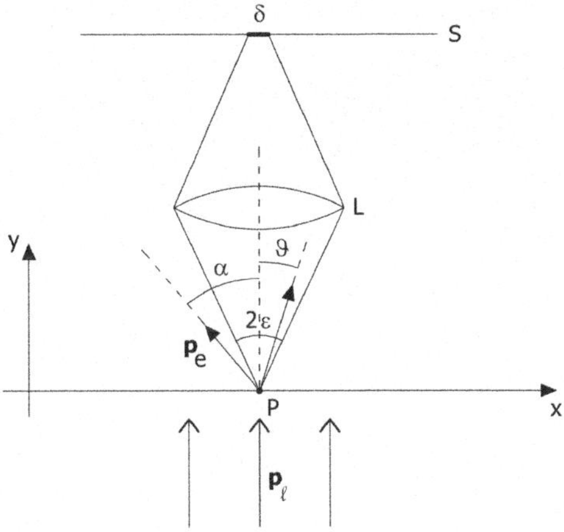

Fig. 3.8 The ideal experiment of Heisenberg's microscope; an electron still in P on the x-axis is illuminated by a beam of light; its image on S through the lens L is a diffraction disk with diameter δ

The Heisenberg principle. A measurement process that allows the determination of a dynamical variable q of a physical system with some accuracy, described by the mean square error (m.s.e.) σ_q, simultaneously puts the system in a state in which the conjugate (in classical sense) variable p has a m.s.e. σ_p satisfying the inequality

$$\sigma_q \sigma_p \geq c, \tag{3.41}$$

where the constant c is of the order of magnitude of h.

Let us notice that this is a principle of the new quantum physics: as such one cannot "prove" it, but it could only be disproved by experience.

This, however, has never happened, so we will consider it one of the pillars on which we are going to build a systematic theoretical body as it has been fixed by P.A.M. Dirac, summarizing all the experiences of the first quantum era as well as the formulations developed independently by Heisenberg and by Schrödinger.

References

1. Caldirola P (1966) Istituzioni di fisica, teorica edn. Viscontea, Milano
2. de Broglie (1951) Onde, corpuscoli, meccanica ondulatoria. Bompiani, Modena
3. Heisenberg (1930) The physical principles of the quantum theory. The University of Chicago Press, Chicago, Translated by C, Eckart and F. C. Hoyt

Chapter 4
The Principles of Quantum Mechanics

4.1 Introduction

From the vehement development of experimental and theoretical physics at the beginning of the twentieth century, we come out with three basic concepts:

(a) *existence of energy quanta*, namely even the electromagnetic radiation can have a "corpuscular" behaviour concentrated in a limited region of space, as if constituted by a flow of particles, photons, with energy $E = h\nu$, and this can be exchanged with atoms in stable status only by packets multiple of a small quantum unit,

(b) *existence of matter waves*, namely even particles of matter, in certain states have a behaviour dictated by the propagation of an associated ψ wave, the square modulus of which is proportional to the probability density of the presence of the particle in a volume element,

(c) *the uncertainty principle*, namely if in a certain dynamic system a variable q is associated with a wave $\psi(q)$ and the conjugate dynamic variable p with a wave $\varphi(p)$, such that

$$\mu_q = E\{q\} = \int q|\psi(q)|^2 dq \ , \ \sigma_q^2 = \int (q - \mu_q)^2 |\psi(q)|^2 dq \ , \qquad (4.1)$$

$$\mu_p = E\{p\} = \int p|\varphi(p)|^2 dp \ , \ \sigma_p^2 = \int (p - \mu_p)^2 |\varphi(p)|^2 dp \ , \qquad (4.2)$$

then the physical state of the system must be such that

$$\sigma_p \sigma_q \geq O(h) \ , \qquad (4.3)$$

so that $\psi(q)$ and $\varphi(p)$ are not arbitrary, but must depend from one another in such a way that (4.3) be satisfied.

F. Sansò and F. Migliaccio, *Quantum Measurement of Gravity for Geodesists and Geophysicists*, Springer Geophysics,
https://doi.org/10.1007/978-3-030-42838-9_4

As we see, the concepts that we are going to fix emerge so to say from the classical physics, like the use of frequencies, impulses, energy, conservation laws, etc.; but the three principles above summarized require that we introduce a new definition of dynamic variables that imply (4.3). Let us underline that by necessity the use of probability here must be much more intrinsic to the physical mechanism, contrary to the classical physics where it has been mostly used to describe the inconsistency of repeated measurements of the same variable, namely the measurement errors, or the overall behaviour of an aggregate of particles. Here the probability distribution becomes the tool to describe the physical state of the system, even when this is constituted by a single particle.

To proceed along this road, we first need a theorem on a functional property of the Fourier transform.

Although not exactly standard, the outlay of this chapter has benefitted from the ideas of Caldirola et al. [1], Dirac [2], Landau and Lifšic [3], Messiah [4].

4.2 A Theorem on Fourier Transform

We aim at proving the following theorem relative to the one-dimensional Fourier transform.

Theorem 4.1 *Let $\psi(x) \in L^2$ be such that*

$$\int_{-\infty}^{+\infty} |\psi(x)|^2 dx = \int_{-\infty}^{+\infty} \psi^*(x)\psi(x)dx = 1 \; ; \tag{4.4}$$

let further $\widehat{\psi}(s)$ denote the Fourier transform of ψ in the L^2 sense, namely

$$\widehat{\psi}(s) = \int_{-\infty}^{+\infty} e^{-i2\pi sx}\psi(x)dx \; . \tag{4.5}$$

Consider now the two positive functionals

$$A(\psi) = \int_{-\infty}^{+\infty} x^2|\psi(x)|^2 dx \tag{4.6}$$

$$B(\psi) = \int_{-\infty}^{+\infty} s^2|\widehat{\psi}(s)|^2 ds \tag{4.7}$$

and also

$$\Phi(\psi) = A(\psi)B(\psi) \; . \tag{4.8}$$

Then $\Phi(\psi)$ has a minimum value on the sphere $\| \psi \|_{L^2} = 1$, i.e. when ψ satisfies (4.4); more precisely it is

$$\min_{\|\psi\|_{L^2}=1} \Phi(\psi) = \frac{1}{16\pi^2} \tag{4.9}$$

and the minimum is achieved at any $\psi(x)$ in the class of functions

$$\forall\sigma > 0\,, \quad \overline{\psi}(x,\sigma) = \frac{1}{\sqrt{\sqrt{2\pi}\sigma}} e^{-\frac{1}{4}\frac{x^2}{\sigma^2}}\,. \tag{4.10}$$

Proof First of all, let us notice that, by the inverse Fourier Transform,

$$\psi(x) = \int_{-\infty}^{+\infty} e^{i2\pi sx}\widehat{\psi}(s)ds\,,$$

so that we can define a linear operator C such that

$$C(\psi) = \frac{1}{2\pi i}\frac{d}{dx}\psi = \int_{-\infty}^{+\infty} e^{i2\pi sx} s\widehat{\psi}(s)ds\,; \tag{4.11}$$

therefore we can write, also recalling Parseval identity,

$$\begin{aligned} B(\psi) &= \int_{-\infty}^{+\infty} |s\widehat{\psi}(q)|^2 dq = \\ &= \int_{-\infty}^{+\infty} |C(\psi)|^2 dx\,. \end{aligned} \tag{4.12}$$

Now, writing the Schwartz inequality for $x\psi(x)$ and $C[\psi(x)]$, recalling too that here we use the complex L^2 space, we find

$$\begin{aligned} &\left|\int_{-\infty}^{+\infty} (x\psi(x)^*)\left(\frac{1}{2\pi i}\frac{d}{dx}\psi(x)\right)dx\right|^2 \leq \\ &\leq \int_{-\infty}^{+\infty} |x\psi(x)|^2 dx \int_{-\infty}^{+\infty} |C(\psi)|^2 dx = A(\psi)B(\psi)\,. \end{aligned} \tag{4.13}$$

Moreover, from the theory of Schwartz inequality, we know that (4.13) can be an equality if and only if

$$\frac{1}{2\pi i}\frac{d}{dx}\psi(x) = cx\psi(x) \tag{4.14}$$

for some complex $c = a + ib$; (4.14) implies

$$\psi(x) = Ke^{\pi icx^2} = Ke^{-b\pi x^2 + ia\pi x^2}\,. \tag{4.15}$$

It is clear that if we want $\psi \in L^2$ we must have $b = Im(c) > 0$; moreover, the existence of $a = Re(c) \neq 0$ modifies $\psi(x)$ by a factor of modulus 1, so it does not change $|\psi|^2$, and we choose to put $a = Re(c) = 0$. So we decide to change the definition of c into

$$c = \frac{1}{4\pi} i \frac{1}{\sigma^2} \tag{4.16}$$

and we get

$$\psi(x) = K e^{-\frac{1}{4}\frac{x^2}{\sigma^2}} .$$

The well-known normalization constant of the Gaussian distribution implies that

$$K^2 = \frac{1}{\sqrt{2\pi}\sigma}$$

and we see that, recalling the definition (4.10) of $\overline{\psi}$,

$$\Phi(\psi) \geq \Phi(\overline{\psi}) , \tag{4.17}$$

$\forall \psi$ of L^2-norm equal to one.

The computation of the minimum $\Phi(\overline{\psi})$ now is straightforward. Recalling (4.13), (4.14) and (4.16) one has

$$\begin{aligned} \Phi(\overline{\psi}) = A(\overline{\psi})B(\overline{\psi}) &\equiv \left| \int (x\overline{\psi}^*) \left(\frac{1}{4\pi\sigma^2} x\overline{\psi} \right) dx \right|^2 \\ &\equiv \frac{1}{16\pi^2\sigma^4} \left| \int x^2 \overline{\psi}^2 dx \right|^2 . \end{aligned}$$

On the other hand, $\overline{\psi}^2 = |\overline{\psi}|^2$ is just a normal distribution with average zero and variance σ^2, so

$$\Phi(\overline{\psi}) = \frac{1}{16\pi^2}$$

as we wanted to prove. □

It is important to underline that the value of the minimum (4.9) is independent of σ^2, as it should. In other words, the minimum is reached at any function of the family $\overline{\psi}(x, \sigma)$. Let us further notice that $\overline{\psi}$ is such that the following two relations are satisfied:

$$\int x|\overline{\psi}|^2 dx = 0, \tag{4.18}$$

$$\int s|\widehat{\overline{\psi}}|^2 ds \equiv |\int \overline{\psi}^* C\overline{\psi} dx| = 0\,, \tag{4.19}$$

so that the minimum of $\Phi(\psi)$ over the set $\{\psi\ ;\ \|\ \psi\ \|_{L^2}= 1\}$ is the same that we get by restricting the search set to

$$\{\psi\ ;\ \|\ \psi\ \|^2_{L^2}= 1\,, < \psi, x\psi >_{L^2}= 0\,,\ < \psi, C\psi >_{L^2}= 0\}\,. \tag{4.20}$$

Due to the obvious invariance under translation of integrals on the x- and s-axes, we can reformulate our result by saying that, fixing any two constants μ_x, μ_s, and defining

$$\widetilde{\Phi}(\psi) = \int (x-\mu_x)^2|\psi(x-\mu_x)|^2 dx \cdot \int (s-\mu_s)^2|\widehat{\psi}(s-\mu_s)|^2 ds$$

one has

$$\Phi(\psi) = \widetilde{\Phi}(\psi) \geq \widetilde{\Phi}(\overline{\psi}) \equiv \Phi(\overline{\psi}) = \frac{1}{16\pi^2} \tag{4.21}$$

for all ψ such that

$$\begin{cases} \int |\psi(x-\mu_x)|^2 dx \equiv \int |\widehat{\psi}(s-\mu_s)|^2 ds \equiv 1 \\ \int (x-\mu_x)|\psi(x-\mu_x)|^2 dx = 0 \\ \int (s-\mu_s)|\widehat{\psi}(x-\mu_s)|^2 ds = 0\,. \end{cases} \tag{4.22}$$

Since it is clear that due to the first of (4.22), $|\psi(x-\mu_x)|^2$ can be interpreted as a one-dimensional probability density function, and the same for $|\widehat{\psi}(s-\mu_s)|^2$ and furthermore μ_x is the mean of the former distribution and μ_s of the latter, as well as $A(\psi)$ is the variance of the former distribution and $B(\psi)$ the variance of the latter, we arrive at a notable stochastic interpretation of Theorem 4.1.

Corollary 4.1 *Let (X, S) be a two-dimensional random variable, on the plane R^2, possessing finite second order moments, and assume that $|\psi(x)|^2$ is the marginal distribution of X and $|\widehat{\psi}(s)|^2$ the marginal distribution of S; then we have, whatever are the means μ_x, μ_s*

$$\sigma(X) \cdot \sigma(S) \geq \frac{1}{4\pi}\,. \tag{4.23}$$

Equality is reached when the marginals are Gaussian distributions.

Remark 4.1 Let us assume that, considering Corollary 4.1, we transform the random variable S into

$$P = hS \tag{4.24}$$

so that, calling $\varphi(p)$ a probability amplitude for p, we will have

$$|\varphi(p)|^2 dp = |\widehat{\psi}(s)|^2 \frac{dp}{h} , \tag{4.25}$$

namely

$$\varphi(p) = \frac{1}{\sqrt{h}} \widehat{\psi}\left(\frac{p}{h}\right) = \frac{1}{\sqrt{h}} \int_{-\infty}^{+\infty} e^{-i\frac{px}{\hbar}} \psi(x) dx \tag{4.26}$$

where we have put, as customary,

$$\hbar = \frac{h}{2\pi} . \tag{4.27}$$

Indeed we will have

$$\sigma(P) = h\sigma(S) \tag{4.28}$$

so that (4.15) becomes

$$\sigma(X)\sigma(P) \geq \frac{1}{2}\hbar . \tag{4.29}$$

Moreover, since by (4.11) we have associated the operator $\frac{1}{2\pi i}\frac{d}{dx}$ to S, we associate to P the operator

$$P \sim \frac{\hbar}{i}\frac{d}{dx} = -i\hbar\frac{d}{dx} = hC . \tag{4.30}$$

We observe that, since the inverse of (4.26) is

$$\psi(x) = \frac{1}{\sqrt{h}} \int_{-\infty}^{+\infty} e^{i\frac{px}{\hbar}} \varphi(p) dp , \tag{4.31}$$

one can easily prove that, by the rules of Fourier Transform, one has, similarly to (4.12),

$$E\{P^2\} = \int_{-\infty}^{+\infty} p^2 |\varphi(p)|^2 dp = \int_{-\infty}^{+\infty} |hC\psi(x)|^2 dx , \tag{4.32}$$

reconducting in another way to the inequality (4.29).

Remark 4.2 Let us notice that when the physical state is represented by the probability amplitude $\psi(q)$, the variable q is just represented by a multiplication and the conjugate variable p, is represented by the operator

$$q \sim q \cdot \quad p \sim -i\hbar \frac{d}{dq} \,. \tag{4.33}$$

However, when the same physical state is represented by a probability amplitude $\varphi(p)$, since going from (4.26) to (4.31) the sign of the exponent is reversed, we see immediately that the variables q and p are exchanged and the sign inverted, namely

$$p \sim p \cdot \quad q \sim i\hbar \frac{d}{dp} \,. \tag{4.34}$$

We conclude this section underlining that although proves are given here in R^1, the same holds in R^3, with suitable changes.

4.3 Some Facts About Hermitian Operators in Hilbert Spaces

Before we can continue in the construction of our quantum model, we need to account for a formalism developed by Dirac [2] in theoretical physics, which parallels but does not coincide with a mathematical development in understanding the spectral nature of Hermitian operators, A, in Hilbert spaces.

The starting point is the definition of spectrum: let us consider the operator $A - \lambda I$ for λ in the complex plane.

The operator

$$R_\lambda = (A - \lambda I)^{-1} \tag{4.35}$$

when it exists and is defined on the whole space (L^2 in our case) is called the resolvent of A. The set of λ values for which R_λ exists is called the resolvent set; one can prove (see [5, VIII, 1]) that this set, $\rho(A)$, is open in C, the complex field.

The complementary set

$$\sigma(A) = \rho(A)^c \tag{4.36}$$

is closed and is called the spectrum.

The following theorem holds (see [5, XI, 8]).

Theorem 4.2 *Every Hermitian operator A on a Hilbert Space H has a real spectrum, $\sigma(A) \subset R$; $\sigma(A)$ can be decomposed into two sets*

$$\sigma(A) = \sigma_d(A) + \sigma_c(A)$$

with $\sigma_d(A) \equiv \{\lambda_n \,;\, n = 1\ldots\}$ *a discrete set and* $\sigma_c(A)$ *which is continuous.*

The values λ_n *of the discrete spectrum are the eigenvalues of* A *and to each of them corresponds at least a vector* $u_n \in H$ *such that*

$$Au_n = \lambda_n u_n \;; \tag{4.37}$$

in general to each λ_n *can correspond an invariant linear subspace* H_n *such that*

$$Au = \lambda_n u \quad u \in H_n \;;$$

two different invariant subspaces are orthogonal

$$< u, v >= 0 \;,\; u \in H_n \;,\; v \in H_m \quad n \neq m \tag{4.38}$$

so that if we identify an orthonormal basis for each subspace

$$\begin{aligned} &\{u_{nm}\} \in H_n \;;\; < u_{nm}, u_{nj} >= \delta_{mj} \\ &Span\{u_{nm}\} \equiv H_n \;, \end{aligned} \tag{4.39}$$

we come, maybe by repeating λ_n *more times, to define a double sequence* $\{\lambda_n, u_n\}$*, with* u_n *orthonormal;* u_n *is then the eigenfunction (eigenvector) of* A *corresponding to the eigenvalue* λ_n*.*

Corollary 4.2 *When the Hermitian operator* A *has only a discrete spectrum, the following spectral decomposition holds:*

$$\begin{cases} A\cdot = \sum_n \lambda_n P_n \\ P_n\cdot = u_n < u_n \,, \cdot > \,, \end{cases} \tag{4.40}$$

where P_n *is the orthogonal projector on* u_n*. If the domain of* A *is dense in* H *one can even prove that* $\{u_n\}$ *is orthonormal and complete.*

Finally, we are interested in the following interpretation of (4.40), because this will guide us to the case of the continuous spectrum.

Define an operator $\mathscr{U} : L^2 \to \ell^2$ (remember that ℓ^2 is the space of complex sequences $c \equiv \{c_n\}$ with norm $\| \{c_n\} \|_\ell^2 = \sum |c_n|^2$)

$$\mathscr{U}\psi \equiv \{c_n\} = \{< u_n, \psi >\} \;;\; \psi \in L^2 \,. \tag{4.41}$$

By Parseval identity we have

$$\| \{c_n\} \|_{\ell^2}^2 = \sum_n |c_n|^2 = \| \psi \|_{L^2}^2 \,, \tag{4.42}$$

namely $\mathscr{U}$ is an isometric isomorphism.

Let us put $\lambda_n = a_n$ an call $a = \{a_n\}$; then we can define an operator $a\cdot$ on ℓ^2 by

$$(a\cdot)c = \{a_n c_n\} . \tag{4.43}$$

Note that in general $(a\cdot)$, as A, is not a bounded operator, so (4.43) might be defined on a subspace of ℓ^2 only. Clearly $(a\cdot)$ is just the multiplication by a_n and it can be thought of as a diagonal (infinite) matrix.

Now, given $c \equiv \{c_n\} \in \ell^2$, we can define the adjoint of $\mathscr{U}$ as

$$\mathscr{U}^\dagger : \ell^2 \to L^2 \; ; \; \mathscr{U}^\dagger c = \sum c_n u_n . \tag{4.44}$$

So $\mathscr{U}^\dagger$ is the adjoint of the isometry $\mathscr{U}$, and since $\{u_n\}$ is orthonormal and complete, we have indeed

$$\mathscr{U}^\dagger \mathscr{U} = I(L^2) \; ; \; \mathscr{U}\mathscr{U}^\dagger = I(\ell^2) \equiv \{\delta_{ik}\} . \tag{4.45}$$

With the above definition, we see that (4.40) can be written as (recall that we have identified $\lambda_n = a_n$)

$$A\cdot = \mathscr{U}^\dagger (a\cdot) \mathscr{U} \; ; \tag{4.46}$$

in other words the operator $\mathscr{U}$ constructed only with the sequence of eigenfunctions $\{u_n\}$ is just the isometric operator that diagonalizes the operator A.

We need now to generalize the above to the case that A has a continuous spectrum, namely a continuous set $J \subset R$; we note that it could very well be that $J \equiv R$. So we assume that A has only a continuous spectrum leaving the mixed case to an easy generalization.

The point is that there is a mathematical formulation of (4.46) when A has a continuous spectrum, namely there exists an isometric isomorphism, $\mathscr{U}$, of $L^2(R) \Leftrightarrow L^2(J)$ such that

$$\forall \psi \in L^2(R) , \quad \varphi(a) = \mathscr{U}[\psi(x)] \in L^2(J) \tag{4.47}$$

and

$$\| \varphi(a) \|_{L^2(J)} \equiv \| \psi(x) \|_{L^2(R)} \; ; \tag{4.48}$$

moreover

$$\mathscr{U}^\dagger[\varphi(a)] \equiv \psi(x) \tag{4.49}$$

and

$$A\psi \equiv \mathscr{U}^\dagger[a\varphi(a)] \equiv \mathscr{U}^\dagger(a\cdot)\mathscr{U}\psi \tag{4.50}$$

(see [5, XI, 14]).

However, to continue our formalism, we need one more restrictive hypothesis.

We assume that there is a distribution valued kernel $u(x, a)$ (we consider it as the kernel of $\mathscr{U}^\dagger$ realizing the identity (4.49)) such that

$$\psi(x) = \int_J da\, u(x, a)\varphi(a) \equiv \mathscr{U}^\dagger\varphi\,, \tag{4.51}$$

or dually

$$\varphi(a) = \int_{-\infty}^{+\infty} dx\, u^*(x, a)\psi(x) \equiv \mathscr{U}\psi \tag{4.52}$$

and, moreover,

$$A\psi(x) \equiv \int_J da\, u(x, a)a \int_{-\infty}^{+\infty} dy\, u^*(y, a)\psi(y)\,. \tag{4.53}$$

Remark 4.3 Notice that, as some elementary examples prove, we need to assume that $u(x, a)$ is in fact a distribution; this is clear if, for instance, we assume that $\mathscr{U} = \mathscr{U}^\dagger \equiv I$. Yet if the relation (4.47) holds for a set of smooth functions (like, for instance, the space $\mathscr{E}$ of fast-decreasing C^∞ functions) which is also dense in L^2, the relations (4.51), (4.52) can be extended by continuity to the whole L^2; so, for instance we can correctly write, after a limit process,

$$\psi(a) = \int_{-\infty}^{+\infty} \delta(x - a)\psi(x)dx$$

$\forall\psi \in L^2(-\infty, +\infty)$.

Given that formally

$$A\psi(x) = \int_J da\, Au(x, a)\varphi(a) \equiv \int_J da\, au(x, a)\varphi(a) \tag{4.54}$$

$\forall\varphi \in L^2(J)$, we represent (4.54) by saying that

$$Au(x, a) = au(x, a) \tag{4.55}$$

and by definition we call a an improper eigenvalue of A and $u(x, a)$ an improper eigenfunction of A.

We prove by two examples that the fundamental dynamic variables q, p represented by the Hermitian operators $q \sim x\cdot,\ p \sim -i\hbar\frac{d}{dx}$, do satisfy our hypothesis on the representation of $\mathscr{U}^\dagger$ by some kernel $u(x, a)$.

Example 4.1 Let q be the operator

$$q\psi(x) \equiv x \cdot \psi(x)\ ;$$

then the continuous spectrum is $J \equiv R$ and the kernel associated to the improper eigenvalue a is just

$$u(x, a) \equiv \delta(x - a)\ .$$

In fact

$$xu(x, a) = x\delta(x - a) \equiv a\delta(x - a)\ ,$$

because for any test function $f(x)$

$$\int xf(x)\delta(x - a)dx \equiv af(a) \equiv a \int \delta(x - a)f(x)dx\ .$$

Example 4.2 Let $p = -i\hbar\frac{d}{dx}$ be the operator representing the momentum in one dimension; then $J \equiv R$ and the improper eigenfunction corresponding to the eigenvalue k is

$$u(x, k) = e^{i\frac{kx}{\hbar}}\ .$$

In fact

$$-i\hbar\frac{d}{dx}e^{i\frac{kx}{\hbar}} \equiv ke^{i\frac{kx}{\hbar}}\ .$$

It is important to observe that the use of any solution of (4.55) is not in general supplying the right kernel of $\mathscr{U}^\dagger$ such that

$$\mathscr{U}^\dagger\mathscr{U} = I(\text{in } L^2(R))\ ,\ \mathscr{U}\mathscr{U}^t = I(\text{in } L^2(J))\ . \tag{4.56}$$

In fact (4.55) is homogenous in $u(x, a)$ and its solution is determined apart from a multiplicative constant, while (4.56) implies that

$$\int da\ u(x, a)u^*(y, a) = \delta(x - y) \quad x, y \in R \tag{4.57}$$

and

$$\int dx\, u(x,a)u^*(x,b) = \delta(a-b) \quad a,b \in J. \tag{4.58}$$

These relations represent normalization conditions, waiving the above ambiguity; it is easy to prove that $u(x,a)$ from Examples 4.1 and 4.2 do satisfy such relations.

At the level of notation however, although we know that $u(\cdot,a) \notin L^2(R)$ and $u(x,\cdot) \notin L^2(J)$, we shall write

$$< u(\cdot,a)\ ,\ u(\cdot,b) >_{L^2(R)} = \delta(a-b)\ , \tag{4.59}$$

and

$$< u(x,\cdot)u(y,\cdot) >_{L^2(J)} = \delta(x-y)\ ; \tag{4.60}$$

this allows to set up a formal rule speeding up calculations.

From the above discussion we understand that it is not possible that for $a \in J$, $u(x,a) \in L^2(R)$, nevertheless we generalize the ordinary concept of orthogonality and completeness in $L^2(R)$ as follows.

Proposition 4.1 *$u(x,a)$ and $u(x,b)$ are "orthogonal", $a \neq b$, meaning that if we take two intervals $\Delta_a, \Delta_b, \Delta_a \cap \Delta_b = \emptyset$ and two L^2 functions φ_1, φ_2 supported respectively on Δ_a and Δ_b one has*

$$\begin{aligned} &< \int_J u(x,a')\varphi_1(a')da'\ ,\ \int_J u(x,b')\varphi_2(b')db' >\equiv \\ &\equiv \int_J \varphi_1^*(a')\varphi_2(a')da' \equiv 0\ . \end{aligned} \tag{4.61}$$

Proposition 4.2 *$\{u(x,a), a \in J\}$ is a complete basis of $L^2(R)$, in the sense that*

$$\begin{aligned} &\int da\, u(y,a) \int u^*(x,a)\psi(x)dx = \\ &= \int dx\, \psi(x) \int da\, u(y,a)u^*(x,a) = \\ &= \int dx\, \psi(x)\cdot\delta(y-x) = \psi(y)\ ; \end{aligned} \tag{4.62}$$

so this is another way to say that $\mathcal{U}^\dagger\mathcal{U} = I$.

Similarly, one has that $\{u(x,a)\ ;\ x \in R\}$ is complete in $L^2(J)$, which is equivalent to state that $\mathcal{U}\mathcal{U}^\dagger = I$.

Remark 4.4 The situation, described separately for discrete and continuous spectrum, is then extended to the case that we have both. In this case Corollary 4.2 and formulas (4.57) and (4.59) are merged to give

$$\sum_n u_n(x)u_n^*(y) + \int_J da\, u(x,a)u^*(y,a) \equiv \delta(x-y)\ ; \tag{4.63}$$

this expresses the completeness of the set of proper and improper eigenfunctions. On the other hand, orthogonality is represented by the relations (using also the notation (4.59)),

$$\begin{cases} < u_n(\cdot), u_m(\cdot) >_{L^2(R)} = \delta_{nm} \\ < u_n(\cdot), u(\cdot, a) >_{L^2(R)} = 0 \\ < u(\cdot, a), u(\cdot, b) >_{L^2(R)} = \delta(a-b)\ . \end{cases} \tag{4.64}$$

The completeness relation (4.63) is also called *closure relation* in the literature on quantum mechanics.

4.4 A More Precise Formulation of Seven Principles of Quantum Mechanics

We can summarize and interpret the discussion of Remark 4.1 by the following proposition: assume that the physical state of a dynamic system is represented by a complex, amplitude function $\psi(x)$, such that $|\psi(\overline{x})|^2$ represents the probability density of the variable X at a value $\overline{x}$, then the canonically conjugated variable p has a probability amplitude $\varphi(p)$ that is related to $\psi(x)$ by the Fourier Transform (4.26) and the variable P itself can be represented by the operator $hC = -i\hbar\frac{d}{dx}$, so that, by using the $\psi(x)$ representation of the state of the system, formulas like

$$E\{P\} = \mu_p =< \psi, hC\psi >\ ,\ E(P^2) =< \psi, (hC)^2\psi > \tag{4.65}$$

hold true. In this way the uncertainty relation (4.29) is verified. This is the reason why we consider our interpretation as a nice mathematical picture of the empirical rule established by Heisenberg in the form of the uncertainty principle.

Generalizing the above considerations, we shall establish seven rules that we can consider as the principles of Quantum Mechanics.

RULE I

Two conjugate variables q, p are represented by operators that depend on the type of amplitude chosen; when $\psi(q)$ is the chosen amplitude we have (see Remark 4.2)

$$q \sim q\cdot\ ,\ p \sim -i\hbar\frac{d}{dq}\ ;$$

when $\varphi(p)$ is the chosen amplitude we have

$$p \sim p \cdot \,, \; q = i\hbar \frac{d}{dp} \,.$$

In any way the two operators are Hermitian, so that their averages given by

$$E\{q\} =< \psi, q\psi > \,, \; E\{p\} =< \psi, -i\hbar \frac{d}{dq}\psi > \tag{4.66}$$

are real. As for $q\cdot$ it is clear that its mean is real, in fact

$$< \psi, q\psi >= \int q|\psi(q)|^2 dq =< q\psi, \psi >=< \psi, q\psi >^* \,.$$

As for p we have (with an integration by parts and denoting by ψ' the derivative of ψ)

$$\begin{aligned} < \psi, -i\hbar\psi' > &= \int \psi^*(-i\hbar\psi')dx = \int i\hbar\psi^{*'}\psi dx = \\ &= < -i\hbar\psi', \psi >=< \psi, -i\hbar\psi' >^* \,. \end{aligned}$$

This generalizes to any couple of canonical conjugate variables Q, P.

RULE II

By direct computation one sees that the commutation rule holds

$$[q, p] = q\left(-i\hbar\frac{d}{dq}\right) - \left(-i\hbar\frac{d}{dq}\right)q\cdot \equiv i\hbar \,; \tag{4.67}$$

but the same commutator is obtained if we use the p-representation (see Rule I). Therefore the commutator of dynamic variables is independent on the representation of the physical state.

RULE III

Any dynamic variable A of a quantum system is represented by a Hermitian operator in L^2 and its mean value, when the system is in the physical state represented by the probability amplitude $\psi \in L^2$, is given by

$$E\{A\} =< \psi, A\psi > \tag{4.68}$$

and is therefore real as it should be.

Let us also underline that when we say that $\psi(x)$ is a probability amplitude we implicitly mean that when ψ represents a physical state it has to be

$$\| \psi \|^2_{L^2(R)} = 1 \,.$$

Since, e.g. for a dynamic system with one degree of freedom we have generally $A = A(p, q)$, there might be ambiguity in constructing the operator A and in particular $A\left(q, -i\hbar\frac{d}{dq}\right)$ might not be Hermitian and so it cannot represent the underlying dynamic variable. The rule chosen in this case is that A is represented by the symmetrized operator

$$A \sim \frac{1}{2}\left[A\left(q, -i\hbar\frac{d}{dq}\right) + A^*\left(q, -i\hbar\frac{d}{dq}\right)\right]. \tag{4.69}$$

For instance, if $A(q, p) = qp$ one has

$$A \sim \frac{1}{2}\left(-i\hbar q\frac{d}{dq} - i\hbar\frac{d}{dq}q\cdot\right) \equiv -qi\hbar\frac{d}{dq} - \frac{1}{2}i\hbar\,.$$

With an integration by parts it is easy to verify that this operator is Hermitian.

RULE IV

Any two canonically conjugate variables (Q, P), represented by two Hermitian operators denoted again Q, P, satisfy the same commutation rule as (q, p), namely

$$[Q, P] = i\hbar\cdot\,, \tag{4.70}$$

i.e. the commutator is invariant under canonical transformations, as the Poisson parentheses were for the classical case. In force of (4.70) Q, P do satisfy the uncertainty relation

$$\sigma(Q)\sigma(P) \geq \frac{1}{2}\hbar\,. \tag{4.71}$$

In fact, notice that $(Q - \mu_Q)$ and $(P - \mu_P)$ do satisfy the same commutation rule as Q, P. Therefore we can assume that Q and P have zero mean, implying

$$\sigma^2(Q) = <\psi, Q^2\psi> = \| Q\psi \|^2\,;\ \sigma^2(P) = <\psi, P^2\psi> = \| P\psi \|^2\,. \tag{4.72}$$

But then, by Schwartz inequality

$$\begin{aligned}\sigma^2(Q)\sigma^2(P) &= \| Q\psi \|^2 \| P\psi \|^2 \geq | < Q\psi, P\psi > |^2 \geq \\ &\geq \left|\frac{1}{2i}\,[< Q\psi, P\psi > - < P\psi, Q\psi >]\right|^2 = \\ &= \frac{1}{4}| < \psi, [QP - PQ]\psi > |^2 = \frac{1}{4}\hbar^2\,,\end{aligned}$$

so proving (4.71).

RULE V

Any dynamic variable A can physically attain only the values of proper or improper eigenvalues; moreover, when (a_n, u_n) are discrete eigenvalues and eigenfunctions of A, the probability that the value of A is a_n, when the physical state of the dynamic system is ψ, is given by

$$P\{A = a_n\} = | < u_n, \psi > |^2 . \tag{4.73}$$

On the other hand, when $(a, u(x, a))$ are continuous eigenvalue and eigenfunction, the probability density that $A = a \in J$ is given by

$$f_A(a|\psi) \equiv | < u(\cdot, a), \psi > |^2 . \tag{4.74}$$

In fact it is enough to write the mean value of A as

$$\begin{aligned} E\{A\} = < \psi, A\psi >= & \Big\langle \sum u_n(x) < u_n, \psi > + \int_J da\, u(x, a) < u(\cdot, a), \psi > , \\ & \sum a_n u_n(x) < u_n, \psi > + \int_J da\, au(x, a) < u(\cdot, a), \psi > \Big\rangle \\ \equiv & \sum a_n | < u_n \psi > |^2 + \int_J da\, a | < u(\cdot, a), \psi > |^2 \end{aligned} \tag{4.75}$$

to recognize that $\{a_n, a \in J\}$ are the possible sample values of A and (4.72), (4.74) represent the mixed, discrete-continuous, probability distribution of A in the state ψ.

RULE VI

From (4.75) we deduce that not only when $\psi = u_n$ or $\psi = u(x, a)$ (this second case is just a mathematical abstraction because $u(x, a)$ cannot be strictly a "physical" state, since it is not in $L^2(\Omega)$), we have that $P(A = a_n|u_n) = 1$ or $f(A = b|u(\cdot, a)) = \delta(a - b)$, but also, vice versa, if we measure A and find either a_n or $a \in J$, then the state ψ of the dynamic system must be either u_n, in the first case, or $u(\cdot, a)$ in the second.

In other words, when we perform a measurement of A we oblige the state of the system to become the eigenfunction corresponding to the observed value of A.

In fact, if

$$\sigma^2(A|\psi) = \sum (a_n - \mu_A)^2 | < u_n, \psi > |^2 + \int_J da(a - \mu_A)^2 | < u(\cdot, a), \psi > |^2 = 0$$

we must also have

$$\begin{aligned} (a_n - \mu_A) < u_n, \psi > &= 0 \\ (a - \mu_A) < u(\cdot, a), \psi) &= 0 . \end{aligned}$$

This can happen either because

$$\mu_A = a_{\overline{n}} \text{ and } < u_n, \psi >= 0 \qquad n \notin \overline{n},$$
$$< u(\cdot, a), \psi >= 0 \,, \; a \in J$$

in which case it is also

$$< u_{\overline{n}}, \psi >= 1 \Rightarrow \psi = u_{\overline{n}} \,,$$

or

$$\mu_A = b \in J \text{ and } < u_n, \psi >= 0 \;\; \forall n$$
$$< u(\cdot, a), \psi >= \delta(a - b)$$

in which case it is also

$$< u(\cdot, a), \psi >= \delta(a - b) \Rightarrow \psi = u(\cdot, b) \,.$$

Let us underline that physically the Rule VI means that a measurement process changes the state ψ of the system obliging it to become either a proper or an improper eigenstate of A.

RULE VII

Two discrete variables A, B are compatible if we can simultaneously determine the value of both of them.

Since when we determine one variable A, the state has to be an eigenfunction of the operator A, if at the same time we determine B we find that the same physical state ψ must be an eigenfunction of both A and B.

Assume for the sake of simplicity that both A, B have only a discrete spectrum $\{a_n\}, \{b_m\}$. Assume further that one of them, e.g. A, has a simple spectrum (i.e. each eigenvalue has one eigenfunction only); then it is clear that $\forall n$, u_n is an eigenfunction of A and B too and so A, B have to commute.

In fact

$$Au_n = a_n u_n \,, \;\; Bu_n = b_n u_n$$

implies

$$ABu_n = a_n b_n u_n = BAu_n \,.$$

If, on the contrary, A has not a simple spectrum but, say, an invariant subspace $H_{n,A}$ corresponds to an eigenvalue a_n, then in any way in $H_{n,A}$ we must be able to find eigenvectors of B, as otherwise we would not be able to determine B exactly.

So, contrary to the above explanation, we will have eigenstates $u_{n,m}$ depending from two quantum numbers (n, m) such that

$$Au_{n,m} = a_n u_{n,m} \ ; \ \ Bu_{n,m} = b_m u_{n,m} \ .$$

Also in this case it is clear that A and B have to commute.

One can prove that, vice versa, given two discrete variables and their Hermitian operators A, B, when they commute they must have a common complete set of eigenfunctions, so that they can always be exactly determined together. We have therefore the following principle

$$(A, B \text{ compatible}) \Leftrightarrow [A, B] = 0 \ . \tag{4.76}$$

Let us notice that if the couples (a_n, b_m) of eigenvalues of (A, B) are all distinct, any other discrete variable, C, compatible with A and B, will be just a function of A, B because for the complete orthonormal sequence $u_{n,m}$ we will have

$$Cu_{n,m} = c_{n,m} u_{n,m} \tag{4.77}$$

so that we can always put

$$F(a_n, b_m) = c_{n,m}$$

to see that $C = F(A, B)$.

In such a case we say that (A, B) form a complete system. Indeed it can happen that the variables forming a complete system for a certain dynamic system are more than two. For instance, if we take a single particle, we know that at least the three Cartesian coordinate (x, y, z) are all commuting with one another and they have also independent eigenfunctions

$$\begin{aligned} xu(x, y, z|x_0) &= x_0 \delta(x - x_0) F(y, z) \\ yu(x, y, z|y_0) &= y_0 \delta(y - y_0) G(x, z) \\ zu(x, y, z|z_0) &= z_0 \delta(z - z_0) H(x, y) \end{aligned}$$

so that the joint system of eigenfunctions will be

$$u(x, y, z|x_0, y_0, z_0) = \delta(x - x_0)\delta(y - y_0)\delta(z - z_0) \ .$$

If the particle has no further internal degrees of freedom, (x, y, z) is a complete system.

4.5 The Time Evolution of a Quantum State: Schrödinger Equation

Let us first remark that a dynamic system constituted by a single particle cannot be described only by one probability distribution of its position $\boldsymbol{x}$ in space, because this will change in time and we need to know how it happens, in other words we need an evolution equation for $\psi(\boldsymbol{x}, t)$. For reasons that will soon be clear, we start with $\psi(\boldsymbol{x}, -t)$. Now we recall that from Example 1.4 we know that $Q = \mathscr{H}$, $P \equiv (-t)$ are canonical conjugate variables. It follows from our Rule IV that the corresponding quantum operators must satisfy the commutation relation

$$[Q, P] = [\mathscr{H}, (-t)] = i\hbar$$

and that, according to Rule I, in a state represented as $\psi(\cdot, P) = \psi(\cdot, -t)$ the operator Q can be represented by

$$Q \equiv \mathscr{H} = i\hbar\frac{d}{dP} = i\hbar\frac{d}{d(-t)} . \tag{4.78}$$

On the other hand, $\mathscr{H} \equiv \mathscr{H}(\boldsymbol{x}, \boldsymbol{p})$ has a precise representation as a Hermitian operator that, according to Rule III is given by

$$H = \frac{1}{2}\{\mathscr{H}(\boldsymbol{x}, -i\hbar\nabla_x) + \mathscr{H}^\dagger(\boldsymbol{x}, -i\hbar\nabla_x)\} . \tag{4.79}$$

So (4.78) applied to the state $\psi(\boldsymbol{x}, -t)$ can be written as

$$i\hbar\frac{d}{d(-t)}\psi(\boldsymbol{x}, -t) \equiv H\psi(\boldsymbol{x}, -t) ,$$

i.e. changing $-t$ into t

$$i\hbar\frac{d}{dt}\psi(\boldsymbol{x}, t) = H\psi(\boldsymbol{x}, t) ; \tag{4.80}$$

Equation (4.80) is Schrödinger equation. As we see, it is an equation of the first order in t so that its solution will be determined in time, when an initial state is given,

$$\psi(\boldsymbol{x}, 0) = \psi_0(\boldsymbol{x}) . \tag{4.81}$$

A first fundamental property that (4.80) has to satisfy is that the "solver" $\mathscr{U}(t)$ of (4.80), namely

$$\psi(\boldsymbol{x}, t) = \mathscr{U}(t)\psi_0(\boldsymbol{x}) , \tag{4.82}$$

must be a unitary operator, so that

$$\begin{aligned} < \psi, \psi > &= < \mathscr{U}(t)\psi_0 \,,\ \mathscr{U}(t)\psi_0 >= \\ &= < \psi_0, \mathscr{U}^\dagger(t)\mathscr{U}(t)\psi_0 >=< \psi_0, \psi_0 >= 1 \,. \end{aligned} \tag{4.83}$$

In fact, if (4.83) were not satisfied, $\psi(\boldsymbol{x}, t)$ could not be a probability amplitude as this implies

$$\int \psi^*(\boldsymbol{x}, t)\psi(\boldsymbol{x}, t)d\boldsymbol{x} \equiv 1 \,. \tag{4.84}$$

The proof of (4.84) stems from the property of H to be Hermitian. In fact from (4.80) we can write, setting $\frac{d}{dt}\psi = \dot{\psi}$,

$$< i\hbar\dot{\psi}, \psi >= -i\hbar < \dot{\psi}, \psi >=< H\psi, \psi >$$

and

$$< \psi, i\hbar\dot{\psi} >= i\hbar < \psi, \dot{\psi} >=< \psi, H\psi > \,.$$

Then, subtracting the two equations and recalling that the last terms are equal, we get

$$i\hbar(< \psi, \dot{\psi} > + < \dot{\psi}, \psi >) = i\hbar\frac{d}{dt} < \psi, \psi >\equiv 0 \,.$$

This shows that $\| \psi \|^2$ has to be constant in time, and therefore equal to 1 if ψ_0 is a probability amplitude.

Remark 4.5 When $\mathscr{H}$ and then H do not depend on time one can find the explicit form $\mathscr{U}(t)$, i.e.

$$\mathscr{U}(t) = e^{-i\frac{Ht}{\hbar}} \,. \tag{4.85}$$

In fact clearly we have

$$\mathscr{U}(0) = I \,; \tag{4.86}$$

and moreover the family (4.85) satisfies the semigroup property

$$\mathscr{U}(t)\mathscr{U}(t') = e^{-i\frac{Ht}{\hbar}}e^{-i\frac{Ht'}{\hbar}} = e^{-i\frac{H(t+t')}{\hbar}} = \mathscr{U}(t + t') \,. \tag{4.87}$$

Notice that (4.87) is correct because both $\mathscr{U}(t)$ and $\mathscr{U}(t')$ are functions of the same operator H, and therefore they commute with one another; otherwise in general $e^{A+B} \neq e^A e^B$.

From (4.87) we have two consequences. If we take $t' = -t$ we get

$$\mathscr{U}(t)\mathscr{U}(-t) = I$$

namely

$$\mathscr{U}(-t) = e^{i\frac{Ht}{\hbar}} = \mathscr{U}(t)^{-1}\,.$$

On the other hand

$$\mathscr{U}(-t) = e^{i\frac{Ht}{\hbar}} = [\mathscr{U}(t)]^{\dagger}$$

so that $\mathscr{U}^{\dagger}(t) = \mathscr{U}(t)^{-1}$, i.e. $\mathscr{U}(t)$ is unitary.

Moreover, if we put $t' = t + dt$ we have

$$e^{-i\frac{H(t+dt)}{\hbar}} - e^{-i\frac{Ht}{\hbar}} = e^{-i\frac{Ht}{\hbar}}(e^{-i\frac{H}{\hbar}dt} - 1) \cong$$
$$\cong e^{-i\frac{Ht}{\hbar}}\left(-i\frac{H}{\hbar}dt\right) = -i\frac{H}{\hbar}e^{-i\frac{Ht}{\hbar}}dt$$

i.e.

$$\dot{\mathscr{U}}(t) = -i\frac{H}{\hbar}\mathscr{U}(t)$$

that applied to ψ_0 shows that

$$\psi(t) = \mathscr{U}(t)\psi_0$$

is a solution of (4.80), reducing to ψ_0 for $t = 0$. Let us notice that we have used the property that any function of a Hermitian operator A commutes with the operator itself. Moreover, this formalism cannot be applied when H is also a function of t and $H(t)$, $H(t')$ do not commute with one another.

Example 4.3 We have seen that the time evolution of a quantum state essentially depends on the shape of the Hamiltonian. We therefore want to know as a first example, what is the Hamiltonian of a particle in a scalar field with potential $V(\boldsymbol{x})$. Since we have classically

$$\mathscr{H} = \frac{p^2}{2m} + V(\boldsymbol{x})$$

we must also have

$$
\begin{aligned}
H &= -\frac{\hbar^2 \nabla \cdot \nabla}{2m} + V(\boldsymbol{x}) = \\
&= -\frac{\hbar^2}{2m} \Delta + V(\boldsymbol{x}) \,.
\end{aligned} \tag{4.88}
$$

This operator is clearly already self-adjoint and so it doesn't require any symmetrization. We observe, for further use, that when the potential is central, i.e. it depends only on $|\boldsymbol{x}|$, we can pass to a spherical coordinate system to get

$$
H = -\frac{\hbar^2}{2m} \left(\frac{\partial^2}{\partial r^2} + \frac{2}{r} \frac{\partial}{\partial r} + \frac{1}{r^2} \Delta_\sigma \right) + V(r) \tag{4.89}
$$

with Δ_σ the usual Laplace Beltrami operator,

$$
\Delta_\sigma = \frac{\partial^2}{\partial \vartheta^2} + \operatorname{ctg}\vartheta \frac{\partial}{\partial \vartheta} + \frac{1}{\sin^2 \vartheta} \frac{\partial^2}{\partial \lambda^2} \,. \tag{4.90}
$$

Example 4.4 Since this will be useful in the sequel, we want to give an example of the Hamiltonian of a particle, for instance, an electron, interacting with an electromagnetic wave. With the help of the e.m. potentials (φ, $\boldsymbol{A}$), also recalling Theorem 2.2 and Eq. (2.79), we can write

$$
H = \frac{1}{2m} \left(p^2 - 2\frac{e}{c} \boldsymbol{p} \cdot \boldsymbol{A} + \frac{e^2}{c^2} A^2 \right) - \varphi \,.
$$

To properly formulate the Hamiltonian operator H we need to symmetrize the term $\boldsymbol{p} \cdot \boldsymbol{A}$, namely

$$
\boldsymbol{p} \cdot \boldsymbol{A} \equiv \frac{1}{2} [-i\hbar \nabla \cdot (\ \boldsymbol{A}\ ) - \boldsymbol{A} \cdot i\hbar \nabla] \equiv -\frac{1}{2} i\hbar \nabla \cdot \boldsymbol{A} - i\hbar \boldsymbol{A} \cdot \nabla \,.
$$

Therefore the Hamiltonian becomes

$$
H = \frac{1}{2m} \left[-\hbar^2 \Delta + i\hbar \frac{e}{c} (\nabla \cdot \boldsymbol{A}) + 2i\hbar \frac{e}{c} \boldsymbol{A} \cdot \nabla + \frac{e^2}{c^2} A^2 \right] - \varphi. \tag{4.91}
$$

4.6 Stationary States; Constants of Motion

Let us consider the eigenstates of a Hamiltonian H that does not depend on time; in general, as we will see by examples in the next chapter, H has both a discrete and a continuous spectrum, namely

$$
\begin{cases}
Hu(\boldsymbol{x}, E_n) = E_n u(\boldsymbol{x}, E_n) & u = 1, 2 \ldots \\
Hu(\boldsymbol{x}, E) = E u(\boldsymbol{x}, E) & E \in J \,.
\end{cases} \tag{4.92}
$$

It is clear that such eigenstates are stationary, in the sense that they represent a probability distribution that does not change in time. In fact, by definition of a function of H, one has, when $\psi_0(x)$ is an eigenstate,

$$\begin{cases} \psi(\boldsymbol{x},t) = e^{-i\frac{Ht}{\hbar}} u(\boldsymbol{x}, E_n) = e^{-i\frac{E_n t}{\hbar}} \; u(\boldsymbol{x}, E_n) \\ \psi(\boldsymbol{x},t) = e^{-i\frac{Ht}{\hbar}} u(\boldsymbol{x}, E) = e^{-i\frac{Et}{\hbar}} \; u(\boldsymbol{x}, E)\,, \end{cases} \tag{4.93}$$

depending whether the eigenstate belongs to the discrete or to the continuous spectrum. On the other hand, since $e^{-i\frac{Et}{\hbar}}$ is just a phase factor of modulus one, we have indeed

$$|\psi(\boldsymbol{x},t)|^2 = |e^{-i\frac{Et}{\hbar}}\psi_0(\boldsymbol{x})|^2 = |\psi_0(\boldsymbol{x})|^2\,,$$

i.e. the probability distribution is constant in time.

Remark 4.6 We can notice that the second part of the Eqs. (4.93) provides a useful tool for the computation of $\psi(\boldsymbol{x},t)$ for whatever initial state $\psi_0(\boldsymbol{x})$.

In fact it is enough to write

$$\psi_0(\boldsymbol{x}) = \sum_n u(\boldsymbol{x}, E_n) \int_R dy\; u^*(\boldsymbol{y}, E_n)\psi_0(\boldsymbol{y}) + \int_J dE\; u(\boldsymbol{x}, E) \int_R u^*(\boldsymbol{y}, E)\psi_0(\boldsymbol{y})dy$$

to realize that

$$\begin{aligned} \psi(\boldsymbol{x},t) \equiv e^{-i\frac{Ht}{\hbar}}\psi_0(\boldsymbol{x}) \equiv \int_R dy \Bigg[& \sum_n u(\boldsymbol{x}, E_n)u^*(\boldsymbol{y}, E_n)e^{-i\frac{E_n t}{\hbar}} + \\ & + \int_J dE\; u(\boldsymbol{x}, E)u^*(\boldsymbol{y}, E)e^{-i\frac{Et}{\hbar}} \Bigg] \psi_0(\boldsymbol{y})\,. \end{aligned} \tag{4.94}$$

Another interesting question, typical of a dynamic system, is to find invariant dynamic variables, i.e. conservation laws. In the case of quantum systems we say that a dynamic variable, A, is conserved if its average is constant in time whatever is the initial state ψ_0; namely

$$\begin{cases} <\psi(\boldsymbol{x},t)\,,\; A\psi(\boldsymbol{x},t)>= \text{ constant} \\ \forall\psi_0(\boldsymbol{x}) = \psi(\boldsymbol{x},0)\,. \end{cases} \tag{4.95}$$

Let us assume once more that H does not explicitly depend on time, otherwise not even the energy is conserved; then (4.95) implies

$$\begin{aligned} 0 = i\hbar\frac{d}{dt} <\psi, A\psi> &= < -i\hbar\dot\psi, A\psi> + <\psi, Ai\hbar\dot\psi> \\ &= < -H\psi, A\psi> + <\psi, AH\psi> = <\psi, [AH - HA]\psi> = \\ &= <\psi_0\,,\; e^{i\frac{Ht}{\hbar}}[A, H]e^{-i\frac{Ht}{\hbar}}\psi_0>\,, \quad \forall\psi_0\,. \end{aligned} \tag{4.96}$$

Since $ie^{i\frac{Ht}{\hbar}}[A, H]e^{-i\frac{Ht}{\hbar}}$ is a self-adjoint operator because, using $[A, H]^\dagger = [H, A] = -[A, H]$,

$$\left[ie^{i\frac{Ht}{\hbar}}[A, H]e^{-i\frac{Ht}{\hbar}}\right]^\dagger = (-i)\left[e^{-i\frac{Ht}{\hbar}}\right]^\dagger [A, H]^\dagger \left[e^{i\frac{Ht}{\hbar}}\right]^\dagger = \\ = ie^{i\frac{Ht}{\hbar}}[A, H]e^{-i\frac{Ht}{\hbar}} ,$$

Equation (4.96) implies that

$$ie^{i\frac{Ht}{\hbar}}[A, H]e^{-i\frac{Ht}{\hbar}} \equiv 0 . \tag{4.97}$$

Considering that unitary operators are always invertible, (4.97) gives

$$[A, H] = 0 .$$

We conclude that A is an invariant dynamic variable if it commutes with H.

Example 4.5 Let us consider a particle in a radial scalar potential, with Hamiltonian given by (4.89).

Now let us introduce two operators, the physical meaning of which will be given in the next chapter

$$M^2 = -\hbar^2 \Delta_\sigma , \quad M_z = -i\hbar \frac{\partial}{\partial \lambda} . \tag{4.98}$$

It is easy to see that

$$[M^2, H] = 0 , \quad [M_z, H] = 0 \tag{4.99}$$

as well as

$$[M^2, M_z] = 0 . \tag{4.100}$$

Therefore both M^2, M_z are constants of motion of the particle. At the same time, we can observe that (H, M^2, M_z) form a system of compatible variables. As we have seen (Rule VII), they must possess a set of common eigenfunctions. We will also show that they form a complete set of dynamic variables for the given dynamic system, namely there is only one eigenfunction corresponding to known values of H, M^2, M_z. Here we look only for proper eigenfunction that belongs to $L^2(R^3)$.

We start by noticing that a function $u(r, \sigma, E)$ with known value of the energy has to satisfy the stationary Schrödinger equation

$$-\frac{\hbar^2}{2m}\left(\frac{\partial^2}{\partial r^2} + \frac{2}{r}\frac{\partial}{\partial r} + \frac{1}{r^2}\Delta_\sigma\right)u(r, \sigma, E) + V(r)u(r, \sigma, E) = Eu(r, \sigma, E) . \tag{4.101}$$

As we will see in the example of the hydrogen atom, it is the condition that $u \in L^2(R^3)$ that constrains the energy E to a discrete set of values E_n. On the other hand, it is clear that, since $u \in L(R^3)$, we can always apply the expansion in a series of spherical harmonics

$$u(r, \sigma, E_n) = \sum_{\ell=0}^{+\infty} \sum_{m=-\ell}^{\ell} u_{\ell m}(r, E_n) Y_{\ell m}(\sigma) \,. \tag{4.102}$$

We have only to consider that in the present context it is convenient to use the complex version of spherical harmonics, namely

$$Y_{\ell m} = \frac{1}{\sqrt{4\pi}} \overline{P}_{\ell m}(\vartheta) e^{im\lambda} \,, \tag{4.103}$$

with $\overline{P}_{\ell m}(\vartheta)$ the associated Legendre functions normalized in such a way that

$$\int_Y Y^*_{\ell m} Y_{jk} d\sigma = \delta_{\ell j} \delta_{mk} \,. \tag{4.104}$$

In fact, we know that in this case $Y_{\ell m}$ are precisely joint eigenfunctions of M^2 and M_z, since

$$M^2 Y_{\ell m} = \hbar^2 \ell(\ell+1) Y_{\ell m} \,, \quad i\hbar \frac{\partial}{\partial \lambda} Y_{\ell m} = -\hbar m Y_{\ell m} \,. \tag{4.105}$$

Substituting then (4.102) into (4.101) we see that $u_{\ell m}$ have to satisfy the one-dimensional differential equation

$$-\frac{\hbar^2}{2m}\left(\frac{\partial^2}{\partial r^2} + \frac{2}{r}\frac{\partial}{\partial r}\right) u_{\ell m} - \frac{\hbar^2}{r^2} \ell(\ell+1) u_{\ell m} + V u_{\ell m} = E_n u_{\ell m} \,. \tag{4.106}$$

In fact, we see from (4.106) that in reality the radial functions $u_{\ell m}(r, \sigma, E_n)$ will not depend on the value of m. In any event, we find that if we specify the value of $E = E_n$, of M^2, i.e. ℓ, and of M_z, i.e. m, we have ultimately found one common eigenfunction only, i.e. $u_\ell(\boldsymbol{x}, E_n) Y_{\ell m}(\sigma)$, which proves that (H, M^2, M_z) form a complete set of variables.

References

1. Caldirola P, Cirelli R, Prosperi GM (1982) Introduzione alla fisica teorica. UTET, Torino
2. Dirac PAM (1959) I principi della meccanica quantistica. Boringhieri, Torino
3. Landau LD, Lifšic EM, (1962) Mécanique quantique. Boringhieri, Torino
4. Messiah A (1965) Mécanique quantique 1–2. Dunod, Paris
5. Yoshida K (1995) Functional analysis. Springer, Berlin

Chapter 5
First Applications of Quantum Theory

5.1 Introduction

The piece of quantum theory presented in this chapter is strictly speaking not all necessary for the understanding of the process of a quantum measurement of gravity. For this reason we will keep the exposition in a quite concise form. Yet we like to present the basic results of the mechanics of the hydrogen atom, at least to show how the general theory of Chap. 4 is capable of reproducing the results of Bohr.

Furthermore, with a quite short introduction to the quantum theory of angular momentum, we will show why empirical results, for instance, the Zeeman effect, cannot be understood without introducing one further variable necessary to explain the behaviour of many particles. This variable, that has the same character as the angular momentum, i.e. it is, so to say, an internal angular momentum of the particle, is called the *spin* and is responsible for the very fine structure of the energetic spectrum of atoms, which is one of the characteristics exploited in the process of quantum measurement of gravity.

For this reason we have deemed it useful an introduction to the matter, although a short one.

The prove of quantum theory explained in this chapter has been derived mostly from the books of Caldirola et al. [1], Messiah [2], and Pauli [3].

5.2 The Free Particle

In this section we simply want to show how the relations studied in Sect. 3.6, that were established on the basis of some reasoning by analogy, can now be retrieved from the principles of quantum mechanics. A free particle in space is a dynamic system with Hamiltonian

F. Sansò and F. Migliaccio, *Quantum Measurement of Gravity for Geodesists and Geophysicists*, Springer Geophysics,
https://doi.org/10.1007/978-3-030-42838-9_5

$$H = \frac{p^2}{2m} = -\frac{\hbar^2}{2m}\Delta \,. \tag{5.1}$$

The stationary states of the system will then have the form

$$\psi_{st}(\boldsymbol{x}, t) = e^{-i\frac{Et}{\hbar}} u(\boldsymbol{x}, E) \,, \tag{5.2}$$

with $u(\boldsymbol{x}, E)$ an improper eigenfunction of the Hamiltonian, i.e.

$$-\frac{\hbar^2}{2m}\Delta u(\boldsymbol{x}, E) = E u(\boldsymbol{x}, E) \,. \tag{5.3}$$

So the eigenstates of the system satisfy the Helmholtz equation

$$\Delta u(\boldsymbol{x}, E) + \frac{2mE}{\hbar^2} u(\boldsymbol{x}, E) = 0 \,. \tag{5.4}$$

As we know, for instance, by reducing (5.4) to the one-dimensional case, such an equation has exponentially divergent solutions when $E < 0$; this case, therefore, cannot be normalized as in (4.58) and we conclude that $E < 0$ is not part of the spectrum of H. On the contrary, when $E > 0$, if we put

$$k^2 = 2mE \,,$$

we find oscillating solutions of (5.4), namely

$$\begin{cases} u(\boldsymbol{x}, E) = e^{i\frac{\boldsymbol{p}\cdot\boldsymbol{x}}{\hbar}} \\ |\boldsymbol{p}| = p = k = \sqrt{2mE} \,, \end{cases} \tag{5.5}$$

that represent standing waves in space, polarized in the direction of $\boldsymbol{p}$ and with a wavelength λ satisfying

$$\frac{1}{\lambda} = \frac{h}{p} \text{ or } \lambda = \frac{p}{h} \,. \tag{5.6}$$

On the other hand, it is clear that (5.5) is the three-dimensional generalization of Example 4.2, and such functions are just the eigenfunctions of the quantum operator $\boldsymbol{p}_{\text{op}}$ because

$$\boldsymbol{p}_{\text{op}} e^{i\frac{\boldsymbol{p}\cdot\boldsymbol{x}}{\hbar}} = -i\hbar\nabla e^{i\frac{\boldsymbol{p}\cdot\boldsymbol{x}}{\hbar}} = \boldsymbol{p} e^{i\frac{\boldsymbol{p}\cdot\boldsymbol{x}}{\hbar}} \,. \tag{5.7}$$

Therefore functions (5.5) are better indexed by the (constant) vector $\boldsymbol{p}$, $u(\boldsymbol{x}, \boldsymbol{p})$, rather then by E, that clearly leaves the direction of $\boldsymbol{p}$ undetermined. On the other hand, since in our system H is function of $\boldsymbol{p}$ only, it is not surprising that

$$[H, \boldsymbol{p}] = 0 ;$$

so H should have a common set of (improper) eigenfunctions with $\boldsymbol{p}$, which is precisely (5.5). In any way the time evolution of $u(\boldsymbol{x}, \boldsymbol{p})$ is, according to (4.93),

$$\psi(\boldsymbol{x}, t) = e^{-i\frac{Et}{\hbar}} u(\boldsymbol{x}, \boldsymbol{p}) \equiv e^{-i2\pi\left(\frac{\boldsymbol{p}\cdot\boldsymbol{x}}{h} - \frac{E}{h}t\right)} \tag{5.8}$$

with $E = \frac{p^2}{2m}$. Comparing this expression with (2.25) we see that the matter wave is a plane wave with wavelength as in (5.6) and frequency

$$\nu = \frac{E}{h} \text{ or } E = h\nu . \tag{5.9}$$

As we see, the formulas of Sect. 3.6 are fully recovered.

5.3 The Hydrogen Atom as an Example

We want to study here the structure of the discrete spectrum of an electron in a hydrogen atom, namely we want to find the proper eigenfunctions of the electron in the Coulomb field generated by the nucleus, i.e. by a proton.

Since the proton has a mass about 1800 times that of the electron, at the present level of discussion we can consider the nucleus as still in space and put the origin of the coordinates in it. The classical potential U of the electron in the electric field of the proton is then

$$(e > 0) \quad U(r) = -\frac{e^2}{r} , \tag{5.10}$$

so that the attraction generated by it is

$$\boldsymbol{F} = -\nabla U(r) = -e^2 \frac{\boldsymbol{r}}{r^3} ,$$

see Fig. 5.1.

We start with some qualitative reasoning. Let us consider the bound states that correspond to energies E_n for which eigensolutions, that go to zero for $r \to \infty$, exist (this is the reason why they are bound states). We know by experience that when an electron is in a stable state inside the atom we have to do some work to extract it and separate it from the nucleus; so we expect that the original energy level of the electron was negative, and therefore the discrete spectrum of the atom has to be negative and we can put

$$-E_n = \frac{\hbar^2}{2m} W_n > 0 . \tag{5.11}$$

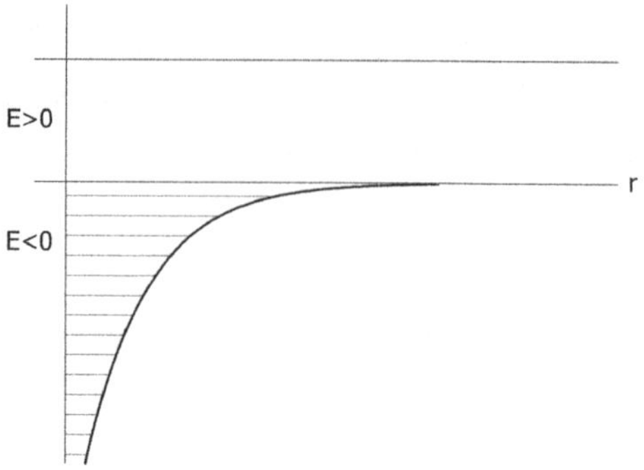

Fig. 5.1 The profile of the Coulomb potential and the spectrum, discrete in $E < 0$ and continuous in $E > 0$

On the contrary, we can easily conceive a situation in which we have a very fast electron that then is not captured by the nucleus, so that it flies away; this requires that the energy of the electron be positive. So we expect that $E > 0$ is the continuous spectrum of the system. This situation is studied in scattering theory and is out of the purpose of the present notes, so we will concentrate only on the eigenvalue equation

$$(H - E_n)u_n(\boldsymbol{x}) = 0. \tag{5.12}$$

Recalling (4.101), this equation is written, introducing W_n, as

$$\frac{\partial^2 u_n}{\partial r^2} + \frac{2}{r}\frac{\partial}{\partial r}u_n + \frac{1}{r^2}\Delta_\sigma u_n + \left[\frac{2m}{\hbar^2}\frac{e^2}{r} - W_n\right]u_n = 0\,. \tag{5.13}$$

Now, using the series representation

$$u_n(r,\sigma) \equiv \sum_{\ell=0}^{+\infty}\sum_{m=-\ell}^{\ell} u_{n\ell}(r)Y_{\ell m}(\sigma) \tag{5.14}$$

we see that (5.13) breaks down into a system of differential equations for the radial functions

$$u''_{n\ell} + \frac{2}{r}u'_{n\ell} + \left[\frac{2me^2}{\hbar^2 r} - W_n - \frac{\ell(\ell+1)}{r^2}\right]u_{n\ell} = 0\,. \tag{5.15}$$

It is convenient to put

$$\frac{me^2}{\hbar^2} = \frac{1}{R} \quad (R \sim 0.5 \cdot 10^{-8}\ \mathrm{cm})$$
$$W_n = k_n^2$$

and write (5.15) as

$$u''_{n\ell} + \frac{2}{r}u'_{n\ell} + \left[\frac{2}{Rr} - k_n^2 - \frac{\ell(\ell+1)}{r^2}\right]u_{n\ell} = 0\,. \tag{5.16}$$

If we look first at the behaviour of the general solution of (5.16) close to the origin, we can observe that in the last coefficient of $u_{n\ell}$, the term $\frac{\ell(\ell+1)}{r^2}$ will be prevailing on the other two, so that, in this region, we expect that $u_{n\ell}$ behaves like a solution of

$$u''_{n\ell} + \frac{2}{r}u'_{n\ell} - \frac{\ell(\ell+1)}{r^2}u_{n\ell} = 0\,.$$

Since this equation has two independent solutions, i.e. $u_{n\ell} \sim r^\ell$, $u_{n\ell} \sim \frac{1}{r^{\ell+1}}$, we guess that the same should happen for (5.16).

On the other hand, we can write

$$\| u_n(r,\sigma) \|^2_{L^2} = \sum_{\ell=0}^{+\infty}(2\ell+1)\int_0^{+\infty} |u_{n\ell}(r)|^2 r^2 dr\,, \tag{5.17}$$

so that only solutions having the behaviour $u_{n\ell} \sim r^\ell$ close to $r = 0$ can make the norm (5.17) finite. The exact solution of (5.16), with the right behaviour at the origin, is given by

$$u_{n\ell}(r) = A_{n\ell} r^\ell e^{-k_n r} F(\ell+1-\nu_n\,|2\ell+2|\,2k_n r), \tag{5.18}$$

where $A_{n\ell}$ is a normalization factor, $F(a, b, x)$ is known as confluent hypergeometric function (see [4, 5]) and

$$\nu_n = \frac{1}{k_n R} = \frac{\sqrt{m}e^2}{\hbar\sqrt{-2E_n}} \tag{5.19}$$

is a non-dimensional parameter. In this formula m, e are the mass and charge of the electron.

The behaviour of the confluent hypergeometric function at infinity is known and given by

$$F(\ell+1-\nu\,|2\ell+2|\,2kr) \underset{r\to\infty}{\sim} \frac{\Gamma(2\ell+2)}{\Gamma(\ell+1-\nu)}(2kr)^{-\ell-1-\nu}e^{2kr}\,, \tag{5.20}$$

yielding, from (5.18).

$$u_{n\ell}(r) \underset{r\to\infty}{\sim} \frac{A_{n\ell}}{(2k)^{\ell+1-\nu_n}} r^{-1-\nu_n} e^{k_n r} \frac{\Gamma(2\ell+2)}{\Gamma(\ell+1-\nu_n)}\,. \tag{5.21}$$

As we see, the behaviour (5.21) will never allow the convergence of (5.17) because of the presence of the exponential term, unless its coefficients in (5.21) are zero. This happens to the inverse Γ function when the argument is zero or a negative integer (see [6]), i.e.

$$\frac{1}{\Gamma(x)} = 0 \Rightarrow x = -n \; ; \; n \geq 0 \, .$$

In our case this becomes, after (5.19),

$$\ell + 1 - \nu_n = -n \; , \; n \geq 0 \Rightarrow E_{n\ell} = -\frac{me^4}{2\hbar^2(\ell + 1 + n)^2} . \qquad (5.22)$$

In other words, the sequence of eigenvalues of the energy is

$$E_N = -\frac{me^4}{2\hbar N^2} \; , \quad N = (\ell + 1 + n) \geq 1 \; , \qquad (5.23)$$

which is identical to Bohr formula (3.20).

To each N correspond quantum numbers

$$0 \leq \ell \leq N - 1 \; , \; -\ell \leq m \leq \ell \, . \qquad (5.24)$$

To each quantum triple (N, ℓ, m) corresponds an eigenstate (stationary state) of the system

$$u_{N,\ell,m}(r, \sigma) = A_{N\ell} r^{\ell} e^{-k_n r} F(\ell - 1 - N \; |2\ell + 2| \; 2k_n r) Y_{\ell m}(\sigma) \qquad (5.25)$$
$$\left(k_N = \frac{me^2}{\hbar^2 N} \right) .$$

Therefore each eigenvalue E_n has a multiplicity equal to

$$\sum_{\ell=0}^{N-1} (2\ell + 1) = N^2 \, . \qquad (5.26)$$

Remark 5.1 The minimum energy state corresponds to $N = 1$, $\ell = 0$, $m = 0$. This eigenvalue is non-degenerate and the relative (unique) eigenfunction has a simple analytic form, namely

$$u_{1,0,0} = Ae^{-\frac{r}{R}} \; ; \qquad (5.27)$$

as we see, the probability distribution is purely radial. The constant A is easily computed form the normalization condition, resulting in the value

$$A = \frac{2}{R^{3/2}} \, . \qquad (5.28)$$

With that one can compute

$$< r >= \frac{4}{R^3} \int_0^{+\infty} e^{-2\frac{r}{R}} r^3 dr = \frac{3}{2} R = 0.75 \cdot 10^{-8} \text{ cm} .$$

This value, that can be considered the standard radius of the hydrogen atom, is confirmed by experience.

5.4 The Quantum Theory of Angular Momentum

This section, together with the next two, is not essential to the understanding of the specific mechanism allowing to perform the quantum measurement of gravity, nor they are at the same high level as the principles of quantum mechanics. Nevertheless, since the measuring procedure as a matter of fact exploits the hyperfine structure of the energy levels of some heavy atoms, we want to give the reader at least the flavour of the physical basis that creates such a complicated structure of these "real" atoms. So, although the argument is interesting and important in itself, we shall proceed rather swiftly without providing many proofs.

The classical angular momentum of a particle is

$$\mathcal{M} = \boldsymbol{r} \wedge \boldsymbol{p} . \tag{5.29}$$

According to our rules the quantum observable angular momentum is

$$\boldsymbol{M} = \boldsymbol{r} \wedge (-i\hbar\nabla) , \tag{5.30}$$

possibly symmetrized, if necessary.

If we express (5.30) in Cartesian coordinates we get

$$\begin{cases} M_x = -i\hbar \left(y\frac{\partial}{\partial z} - z\frac{\partial}{\partial y} \right) \\ M_y = -i\hbar \left(z\frac{\partial}{\partial x} - x\frac{\partial}{\partial z} \right) \\ M_z = -i\hbar \left(x\frac{\partial}{\partial y} - y\frac{\partial}{\partial x} \right) . \end{cases} \tag{5.31}$$

As we see, the products of operators in brackets are commuting (e.g. $x\frac{\partial}{\partial y} = \frac{\partial}{\partial y}x$), and therefore no symmetrization is needed. By direct computation one can verify that the above Cartesian components satisfy the commutation relations

$$[M_x, M_y] = i\hbar M_z , \ [M_y, M_z] = i\hbar M_x , \ [M_z, M_x] = i\hbar M_y . \tag{5.32}$$

For reasons that will be soon become apparent, it is convenient to express $\boldsymbol{M}$ in spherical coordinates.

We have, calling as usual $\sigma = (\vartheta, \lambda)$ the angular coordinates,

$$\begin{aligned} \boldsymbol{M} &= -i\hbar\boldsymbol{r} \wedge \nabla = -i\hbar\boldsymbol{r} \wedge \left(\boldsymbol{e}_r \frac{\partial}{\partial r} + \frac{1}{r}\nabla_\sigma\right) = \\ &= -i\hbar\boldsymbol{e}_r \wedge \left(\boldsymbol{e}_\vartheta \frac{\partial}{\partial \vartheta} + \frac{\boldsymbol{e}_\lambda}{\sin\vartheta}\frac{\partial}{\partial \lambda}\right) = \\ &= -i\hbar\boldsymbol{e}_\lambda \frac{\partial}{\partial \vartheta} - \frac{\boldsymbol{e}_\vartheta}{\sin\vartheta}\frac{\partial}{\partial \lambda}, \end{aligned} \tag{5.33}$$

where $\boldsymbol{e}_r, \boldsymbol{e}_\vartheta, \boldsymbol{e}_\lambda$ are the unit vectors tangent to the r, ϑ, λ lines, respectively.

From (5.33), with some computations, one obtains

$$M^2 = |\boldsymbol{M}|^2 = -\hbar^2\left(\frac{\partial^2}{\partial\vartheta^2} + \mathrm{ctg}\vartheta\frac{\partial}{\partial\vartheta} + \frac{1}{\sin^2\vartheta}\frac{\partial^2}{\partial\lambda^2}\right) = -\hbar^2\Delta_\sigma \tag{5.34}$$

Similarly, one derives

$$\boldsymbol{M}_z = \boldsymbol{e}_z \cdot \boldsymbol{M} = i\hbar\frac{\partial}{\partial\lambda}\,. \tag{5.35}$$

Formulae (5.34) and (5.35) explain the notation introduced in (4.105).

We have already stated that for a particle in the central field the set (H, M^2, M_z) is complete.

Moreover, (5.34) and (5.35) tell us that

$$[M^2, M_z] = 0\,; \tag{5.36}$$

however, since a rotation of the axes will not change M^2 and it can send the z-axis in any (fixed) direction in space, we must have as well

$$[M^2, M_x] = 0\,,\ [M^2, M_y] = 0\,,$$

namely

$$[M^2, \boldsymbol{M}] = 0\,. \tag{5.37}$$

Since M^2 and M_z commute, they have a common set of eigenfunctions and we already know that these are the complex spherical harmonics $\{Y_{\ell m}(\vartheta, \lambda)\}$, $(\ell = 0, 1, 2\ldots, m = -\ell, \ldots 0, \ldots \ell)$, i.e.

$$M^2 Y_{\ell m} = \hbar^2\ell(\ell+1)Y_{\ell m}\,;\ M_z Y_{\ell m} = \hbar m Y_{\ell m}\,. \tag{5.38}$$

It is interesting to notice that, since

$$\max|M_z| = \hbar\ell$$

we always have

$$M_z^2 \leq M^2 = \hbar^2 \ell(\ell + 1)\ ; \tag{5.39}$$

this is logical because the square of a Hermitian operator is non-negative and therefore

$$M^2 - M_z^2 = M_x^2 + M_y^2 \geq 0\ .$$

Remark 5.2 As claimed in our Rule II, the commutation relations between dynamic variables are more intrinsic to a dynamic system than their representation in terms of specific operators. So it is not just an exercise to perform an abstract analysis of the eigenvalues of the couple (M^2, M_z), based only on (5.32) and, consequently, on the Eq. (5.38). The result, that we will not prove and that can be found in [7], can be summarized in two statements:

(a) let a vector $\boldsymbol{M}$ of Hermitian operators satisfy (5.32), then M^2 can attain only values $\hbar^2\ell(\ell + 1)$ with ℓ zero, semi-integer or integer positive number

$$\ell = 0, 1/2, 1, 3/2, 2 \ldots \tag{5.40}$$

(b) given the value $\hbar^2\ell(\ell + 1)$ of M^2, the operator M_z can attain only one of the $(2\ell + 1)$ values $m\hbar$, with

$$m = -\ell, -\ell + 1 \ldots \ell - 1, \ell\ . \tag{5.41}$$

Notice that when ℓ is semi-integer, m can never get the value 0.

So we realize that the (orbital) angular momentum of a particle cannot have a squared modulus covering the full range of "theoretical" values derived directly from (5.36).

This remark has opened the door to the introduction of a new quantum variable of elementary particles, called the spin, that is essential to fully understand their dynamics. We will present it in Sect. 5.6.

5.5 First-Order Spectral Analysis of Static Perturbations of the Hamiltonian

Let's assume we have a dynamic system governed by a Hamiltonian H_0, that does not explicitly depend on time and that we have studied to determine its spectrum. In particular, we suppose to have found only a discrete spectrum $\{E_{on}\}$, such that

$$n = 1, 2 \ldots \quad H_0 u_{on} = E_{on} u_{on}\ . \tag{5.42}$$

Now let's consider another dynamic system that we assume to be close to the previous, in the sense that the new Hamiltonian H is given by

$$H = H_0 + \varepsilon\delta H \tag{5.43}$$

where we know from physical experience that $\varepsilon\delta H$ is small compared to H_0.

The point is whether we can relate the spectrum of H, discrete by hypothesis, to that of H_0. The problem has opened a large chapter of Mathematical Analysis (see [8, 9]), yet we are interested into first-order perturbations, under the hypothesis that they are sufficient to bring us close enough to the exact solution.

We derive the solution to this problem under two different hypotheses: (a) that the spectrum of H_0 is non-degenerate, i.e. for each E_{on}, there is only one u_{on} solving (5.38); (b) that the spectrum of H_0 is multiple but on the other hand that δH commutes with H_0

$$[H_0, \delta H] = 0\,. \tag{5.44}$$

a) Let the spectrum of H_0 be simple; we will concentrate on a specific eigenvalue E_{on} and on its eigenfunction u_{on}. Since the spectrum of H_0 is discrete, we will have a minimum distance between E_{on} and the neighbours E_{on-1}, E_{on+1}. One can prove that if we take ε as a varying parameter and we put $H(\varepsilon) = H_0 + \varepsilon\delta H$, there is an eigenvalue $E_n(\varepsilon)$ of $H(\varepsilon)$ such that, when $\varepsilon \to 0$, we have

$$E_n(\varepsilon) \to E_{on}\,.$$

Then we know that, below a certain $\overline{\varepsilon}$, we have

$$\varepsilon \leq \overline{\varepsilon}\,, \quad E_{on-1} < E_n(\varepsilon) < E_{on+1}\,, \tag{5.45}$$

and at the same time $E_n(\varepsilon)$ will be the closest to E_{on}. We assume that the physical value of ε in (5.43) is in such a situation, so that it can be unambiguously identified.

Then if we write a Taylor expansion of $E_n(\varepsilon)$, $u_n(\varepsilon)$ in the parameter ε, setting

$$E_n(\varepsilon) = E_{on} + \varepsilon E_{1n} + O(\varepsilon^2), \tag{5.46}$$

$$u_n(\varepsilon) = u_{on} + \varepsilon u_{1n} + O(\varepsilon^2)\,, \tag{5.47}$$

we can form the (exact) eigenvalue equation

$$\begin{aligned} Hu_n(\varepsilon) &= (H_0 + \varepsilon\delta H)(u_{on} + \varepsilon u_{1n} + O(\varepsilon^2)) = \\ &= E_n u_n(\varepsilon) = (E_{on} + \varepsilon E_{1n} + O(\varepsilon^2))(u_{on} + \varepsilon u_{1n} + O(\varepsilon^2))\,. \end{aligned} \tag{5.48}$$

Multiplying and equating terms with the same power of ε, stopping at the first order, we get

$$H_0 u_{on} = E_{on} u_{on}, \tag{5.49}$$

$$\delta H u_{on} + H_0 u_{1n} = E_{1n} u_{on} + E_{on} u_{1n} . \tag{5.50}$$

Equation (5.49) is trivially true.
As for (5.50), we develop u_{1n} on the basis $\{u_{0k}\}$, writing

$$u_{1n} = \sum_k a_{nk} u_{0k} , \tag{5.51}$$

and use such expression in it, to find, after a rearrangement,

$$\sum_k (E_{ok} - E_{on}) a_{nk} u_{on} = E_{1n} u_{on} - \delta H u_{on} . \tag{5.52}$$

Writing the scalar product of (5.52) with u_{oj} we get the algebraic equation

$$(E_{oj} - E_{on}) a_{nj} = E_{1n} \delta_{jn} - < u_{oj} , \delta H u_{on} > . \tag{5.53}$$

As we see, setting alternatively $j = n$ and $j = k \neq n$, we find

$$\begin{cases} O = E_{1n} - < u_{on}, \delta H u_{on} > \\ (E_{on} - E_{ok}) a_{nk} = < u_{ok}, \delta H u_{on} > . \end{cases} \tag{5.54}$$

Equations (5.54) say that

$$E_{1n} = < u_{on}, \delta H u_{on} > , \tag{5.55}$$

which by the way is a real number as it should, and

$$k \neq j \qquad a_{nk} = \frac{< u_{ok}, \delta H u_{on} >}{E_{on} - E_{ok}} . \tag{5.56}$$

As we see E_{1n} is determined and, using (5.56) in (5.51), u_{1n} is determined too, but for its component on u_{on}.
If the underlying Hilbert space were real, we could derive the condition of orthogonality of u_{1n} to u_{on} from the requirement that both u_{on} and $u_n(\varepsilon)$ have to be of unitary norm, yet for a complex Hilbert space this condition gives

$$1 = \| u_n \|^2 = \| u_{on} + \varepsilon u_{1n} \|^2 = \| u_{on} \|^2 + \varepsilon [< u_{on}, u_{1n} > + < u_{1n}, u_{on} >] + O(\varepsilon^2) .$$

So, retaining first-order quantities only, we get

$$< u_{on}, u_{1n} > + < u_{1n}, u_{on} > = 2Re < u_{on}, u_{1n} > = 0 . \tag{5.57}$$

Namely $< u_{on}, u_{1n} >= a_{nn}$ has to be a pure imaginary number, but for the rest it is undetermined. Nevertheless we can take the minimum norm u_{1n} satisfying (5.56) and (5.57), maintaining that our solution to the perturbation of the eigenstate is

$$u_{1n} = \sum_{k \neq n} \frac{< u_{ok}, \delta H u_{on} >}{E_{on} - E_{ok}} u_{on} \ . \tag{5.58}$$

Equations (5.55) and (5.58) solve the posed problem.

(b) Now we assume that H_0 has a degenerate spectrum, but at the same time the perturbation δH commutes with H_0. In this case we know that there exists a orthonormal basis of L^2 that is common to both, i.e.

$$\exists\{u_{nm}\} \ : \ \begin{aligned} H_0 u_{nm} &= E_{on} u_{nm} \\ \delta H u_{nm} &= E_{1nm} u_{nm} \ . \end{aligned} \tag{5.59}$$

But in this case

$$E_{1nm} =< u_{nm}, \delta H u_{nm} > \tag{5.60}$$

and

$$H u_{nm} = (E_{onm} + \varepsilon E_{1nm}) u_{nm} \ . \tag{5.61}$$

In other words, $\{u_{nm}\}$ does not change, while the new eigenvalues are determined by (5.60).

We underline that only in case $(H_0, \delta H)$ form a complete system, the sequence $\{u_{nm}\}$ is uniquely identified; nonetheless the above equations hold true.

5.6 The Zeeman Effect, the Spin, and the Atomic Spectrum

The Zeeman effect is a result we obtain by studying the spectral structure of the energy levels of an atom when it is embedded in a constant magnetic field $\boldsymbol{H}$. In fact we know that if we hit an atom with a light beam of well-defined frequency ν we might have an absorption of quanta $h\nu$ corresponding to the jump between two energy levels of the external (so-called optical) electrons, according to the relation

$$E_k - E_n = h\nu. \tag{5.62}$$

In this way, by suitably changing ν, we can experimentally study the spectrum, e.g. of the outermost electron of the atom.

Now suppose for the sake of simplicity that this electron interacts with the rest of the atom by a central Coulomb field $U(r)$, then we know that its energy levels will be generally organized according to two quantum numbers, $\{E_{n\ell}\}$ (see Sect. 5.4). We know as well that each $E_{n\ell}$ corresponds to the value $(E_{n\ell})$ of the energy H and a given value $\hbar^2\ell(\ell+1)$ (with ℓ integer) of the squared angular momentum M^2 (see Sect. 5.6).

Yet to $E_{n\ell}$ in general correspond $(2\ell+1)$ eigenfunctions, common to H, M^2 and to the operator M_z, such that M_z can assume values $\hbar m$, $(|m| \leq \ell)$. The case of the hydrogen atom is exceptional since the energy levels depend on a quantum number only and the degeneracy is even larger. Yet, let us assume that we perturb our dynamic system by introducing an operator δH proportional to M_z.

In this case indeed, as discussed in Sect. 5.5, (b), we expect that the eigenstates are not perturbed, but the energy levels of the new system will become dependent on three quantum numbers, namely $E_{n,\ell,m}$. In other words, we expect that each line of energy $E_{n\ell}$ (or frequency $\nu_{n\ell}$) is split into $(2\ell+1)$ lines $E_{n,\ell,m}$.

Experiments conducted by Pieter Zeeman, that won him the Nobel prize in 1902, showed that, contrary to the previous theory, one obtains $2(2\ell+1)$ spectral lines as opposed to the expected $(2\ell+1)$.

First of all, let us see what is the theoretical frame of the experiment.

An electron in a central field has Hamiltonian

$$H_0 = \frac{1}{2m}p^2 + U(r) \,. \tag{5.63}$$

Now let us add the presence of a static magnetic field, for instance, directed along the z-axis. Let's call $H_m \boldsymbol{e}_z$ such magnetic field and notice that, if you put

$$\boldsymbol{A} = \frac{1}{2}H_m \boldsymbol{e}_z \wedge \boldsymbol{r} \tag{5.64}$$

you get its vector potential, since $\dot{\boldsymbol{A}} = 0$ and

$$\nabla \wedge \boldsymbol{A} = H_m \boldsymbol{e}_z \,, \tag{5.65}$$

as one easily verifies, by a direct computation of the components.

Therefore, referring to (2.79), we have the new Hamiltonian in the form

$$\begin{aligned} H &= \frac{1}{2m}p^2 - \frac{e}{cm}(\boldsymbol{p}\cdot\boldsymbol{A})^s + \frac{e^2}{2mc^2}A^2 + U(r) \equiv \\ &\equiv H_0 - \frac{e}{cm}(\boldsymbol{p}\cdot\boldsymbol{A})^s + \frac{e^2}{2mc^2}A^2, \end{aligned} \tag{5.66}$$

where the index s refers to our Rule III dictating the need of symmetrizing any operator representing an observable. The last term in (5.66) is negligible, the second term on the contrary is our $\varepsilon\delta H$ treated in Sect. 5.6. However, as we see, the operator needs to be symmetrized. Namely

$$\begin{aligned}(\boldsymbol{p}\cdot\boldsymbol{A})^s &= \frac{1}{2}[-i\hbar\nabla\cdot(A\ )-i\hbar\boldsymbol{A}\cdot\nabla]\\ &\equiv \frac{1}{2}[-i\hbar(\nabla\cdot\boldsymbol{A})\cdot-2i\hbar\boldsymbol{A}\cdot\nabla].\end{aligned}$$

Let us notice that for $\boldsymbol{A}$ as in (5.64) one has $\nabla\cdot\boldsymbol{A}=0$ because $\boldsymbol{e}_z$ is constant and $\boldsymbol{r}=\frac{1}{2}\nabla r^2$.

So we obtain

$$\begin{aligned}\delta H &= i\frac{e\hbar}{cm}\boldsymbol{A}\cdot\nabla=\frac{e\hbar H_m}{2cm}\boldsymbol{e}_z\wedge\boldsymbol{r}\cdot i\nabla\equiv\\ &\equiv H_m\frac{e\hbar}{2cm}\boldsymbol{e}_z\cdot\boldsymbol{r}\wedge i\nabla=-H_m\frac{e}{2cm}M_z.\end{aligned}\tag{5.67}$$

So we are perfectly in the situation described at the end of Sect. 5.6 and we see that δH expresses the coupling of the external magnetic field with that generated by the orbital momentum. However the Zeeman effect experimentally shows that each energy line, related to a specific value of M_z, is further split into two.

Things happen as if the electron had an additional degree of freedom of the kind magnetic momentum, that couples with the external magnetic field.

Based on this intuition and on the fact that the orbital momentum could explain only the integer values but not the half integers, W. Pauli proposed a theory based on the idea that the electron possesses an internal angular momentum, with the z component S_z capable of attaining the values $\pm\frac{1}{2}\hbar$ and accordingly also a magnetic momentum $\boldsymbol{\mu}$ proportional to S_z. So the new perturbation term of the Hamiltonian gets the form

$$\delta H=-\frac{eH_m}{2cm}M_z-\mu H_m S_z\tag{5.68}$$

with μ a suitable constant.

As for this new quantum variable, i.e. the spin $\boldsymbol{S}$ of the electron, we will postulate that its components have to satisfy the commutation relations typical of the angular momentum (5.32), i.e.

$$[S_x,S_y]=i\hbar S_z\ ,\ [S_y,S_z]=i\hbar S_x\ ,\ [S_z,S_x]=i\hbar S_y\ .\tag{5.69}$$

To determine the shape of the operators S_x, S_y, S_z it is convenient to switch to the so-called *Pauli matrices*, namely we put

$$\boldsymbol{S}=\frac{1}{2}\hbar\boldsymbol{\sigma}\tag{5.70}$$

so that (5.69) becomes simply

$$[\sigma_x, \sigma_y] = 2i\sigma_z\ , \ [\sigma_y, \sigma_z] = 2i\sigma_x\ , \ [\sigma_z, \sigma_x] = 2i\sigma_y\ . \tag{5.71}$$

Then we use a specific representation in which the operator σ_z is diagonal and, as we know, has eigenvalues $+1, -1$; this corresponds to the fact that S_z attains the values $\pm\frac{1}{2}\hbar$, as we derive from experiments. Therefore σ_z will be a 2×2 matrix, precisely

$$\sigma_z = \begin{vmatrix} 1 & 0 \\ 0 & -1 \end{vmatrix}\ . \tag{5.72}$$

We note explicitly that (5.72) implies

$$\sigma_z^2 = \begin{vmatrix} 1 & 0 \\ 0 & 1 \end{vmatrix} = I \tag{5.73}$$

Substituting σ_x from the second of (5.71) into the first, we derive the remarkable relation

$$\sigma_z\sigma_y\sigma_z = -\sigma_y\ ,$$

that, multiplied to the right by σ_z gives, using (5.73),

$$\sigma_z\sigma_y = -\sigma_y\sigma_z\ , \tag{5.74}$$

i.e. σ_y and σ_z anti-commute. The same is true for any couple of components of $\boldsymbol{\sigma}$. Taking such property into account, we transform the commutation relation into

$$\begin{cases} \sigma_x\sigma_y = i\sigma_z \\ \sigma_y\sigma_z = i\sigma_x \\ \sigma_z\sigma_x = i\sigma_y\ . \end{cases} \tag{5.75}$$

Now let us put

$$\sigma_x = \begin{vmatrix} a & b \\ b^* & c \end{vmatrix}\ , \ \sigma_y = \begin{vmatrix} e & f \\ f^* & g \end{vmatrix} \tag{5.76}$$

so that σ_x, σ_y are self-adjoint matrices.

By explicitating the relation (5.74) we immediately get

$$e = g = 0\ . \tag{5.77}$$

From a similar relation for σ_x, σ_z we get as well

$$a = c = 0\,. \tag{5.78}$$

If we then use (5.76), with zero on the diagonals, into the last two of (5.75), we get

$$f = -ib\,. \tag{5.79}$$

Finally by using the first of (5.75) we find

$$i|b|^2 = i\,.$$

By choosing the solution $b = 1$ we also find $f = -i$, i.e.

$$\sigma_x = \begin{vmatrix} 0 & 1 \\ 1 & 0 \end{vmatrix}\,, \quad \sigma_y = \begin{vmatrix} 0 & -i \\ i & 0 \end{vmatrix} \tag{5.80}$$

In this way we come to specify the operators for the spin of the electron. It turns out that many elementary particles have spin $\frac{1}{2}$ (this means that each component can only have values $\pm\frac{1}{2}\hbar$): for instance, protons and neutrons. But there are other particles that happen to possess a spin equal to 1, for which a similar analysis can be conducted. In this case, the σ_z matrix is determined by the requirement that it must be diagonal and have eigenvalues $+1, 0, -1$ (corresponding to $S_z = \frac{1}{2}\hbar, 0, -\frac{1}{2}\hbar$), i.e.

$$\sigma_z = \begin{vmatrix} 1 & 0 & 0 \\ 0 & 0 & 0 \\ 0 & 0 & 1 \end{vmatrix}$$

Higher values of spin are possible too, but in general we have particles with half integer spin, called *Fermions*, and particles with integer spin, called *Bosons*.

Fermions and Bosons, when collected in an assembly of N elements, display a quite different physical behaviour. But, since this goes beyond the theory we will need in the next chapter, we stop here considering only particles with spin $\frac{1}{2}$, because such are the basic constituents of the atoms and this allows us to introduce the next important remark on the qualitative form of atomic spectra.

Remark 5.3 The study of the Zeeman effect essentially consists in the determination of the perturbation of the energy spectrum of an atom embedded in a constant magnetic field. Such perturbation is caused by a change in Hamiltonian, introducing a new term for the interaction between the external magnetic field, the magnetic field created by the orbital angular momentum and the intrinsic magnetic field of the electron related to the spin. Yet one can immediately imagine, on the basis of classical analogies, that there must be an energy due to the coupling of the orbital magnetic momentum and that of the electron, and one even due to the coupling of the magnetic moment of the electron with that of the protons contained into the nucleons.

Since the effects have different, decreasing order of magnitude, one can study them by repeatedly applying the perturbation formulas. The result is a very complicated structure of the energy level of optical electrons, where each original line, determined by taking into account only a Coulomb interaction, is broken into fine and hyperfine multiplets representing the above interactions.

5.7 Time-Dependent Perturbations of the Hamiltonian

To close the chapter we add a piece of perturbation theory that we deem introductory to our matter. We consider a dynamic system characterized by a Hamiltonian of the shape

$$H = H_0 + \varepsilon \delta H(t), \tag{5.81}$$

where H_0 is independent of time and ε is a small positive constant, in the sense that

$$\forall \psi \, , \quad \varepsilon | < \psi, \delta H(t) \psi > | \ll | < \psi, H_0 \psi > | \, . \tag{5.82}$$

We underline that in general $\delta H(t)$ does not commute with H_0, nor $\delta H(t), \delta H(t')$ commute with one another when $t \neq t'$.

Our purpose is to find at least a perturbative solution of the Schrödinger equation

$$\begin{cases} i\hbar\dot{\psi}(t) = (H_0 + \varepsilon\delta H(t))\psi(t) \\ \psi(0) = f \, . \end{cases} \tag{5.83}$$

Note that in (5.83) and in subsequent formulas we make explicit the dependence of the probability amplitude ψ on time and not on other variables.

We then put

$$\psi(t) = \psi_0(t) + \varepsilon\delta\psi(t) + O(\varepsilon^2) \tag{5.84}$$

and substitute in (5.83).

Separating zero- and first-order terms in ε and neglecting higher order terms, we get the perturbative scheme

$$\begin{cases} i\hbar\dot{\psi}_0(t) = H_0\psi_0(t) \\ \psi_0(0) = f \, , \end{cases} \tag{5.85}$$

$$\begin{cases} i\hbar\delta\dot{\psi}(t) = H_0\delta\psi(t) + \delta H(t)\psi_0(t) \\ \delta\psi(0) = 0 \, . \end{cases} \tag{5.86}$$

Let us notice that there is a certain freedom in distributing the initial value f or part of it between ψ_0 and $\varepsilon\delta\psi$, yet the above choice is clearly the simplest.

The solution of (5.85) has been given in Remark 4.5, and writes

$$\psi_0(t) = e^{-i\frac{H_0 t}{\hbar}} f \, . \tag{5.87}$$

The solution of (5.86) can be written as

$$\delta\psi(t) = -\frac{1}{\hbar}\int_0^t e^{-i\frac{H_0(t-\tau)}{\hbar}} \delta H(\tau)\psi_0(\tau) d\tau \, . \tag{5.88}$$

This can be directly verified, taking into account that

$$i\hbar\frac{d}{dt}e^{-i\frac{H_0(t-\tau)}{\hbar}} = H_0 e^{-i\frac{H_0(t-\tau)}{\hbar}} \tag{5.89}$$

and that

$$e^{i\frac{H_0 \cdot O}{\hbar}} \equiv I \, . \tag{5.90}$$

It is also obvious that $\delta\psi(0) = 0$.

Let us notice that, substituting (5.87) into (5.88), one can write

$$\delta\psi(t) = -\frac{i}{\hbar}\int_0^t e^{-i\frac{H_0(t-\tau)}{\hbar}} \delta H(\tau) e^{-i\frac{H_0\tau}{\hbar}} f d\tau \, , \tag{5.91}$$

and this gives rise to the perturbed solution

$$\psi(t) = e^{-i\frac{H_0 t}{\hbar}} f - \frac{i\varepsilon}{\hbar}\int_0^t e^{-i\frac{H_0(t-\tau)}{\hbar}} \delta H(\tau) e^{-i\frac{H_0\tau}{\hbar}} f d\tau \, . \tag{5.92}$$

Remark 5.4 One interesting application of (5.92) is the calculation of the so-called *transition probability* between two bound states of H_0.

Assume that H_0 has a discrete spectrum $\{E_{oj}\}$, with eigenstates $\{u_{oj}\}$, corresponding to bound states ($u_{oj} \in L^2$). Assume further that the system is in the state u_{oj} at time $t = 0$.

Now we consider the case that the perturbation $\varepsilon\delta H(t)$ has a finite time extent, namely it is switched on at $t = 0$ and switched off at $t = T$. It is interesting to know what is the probability that, under the given perturbation, the system has moved from u_{oj} to u_{ok}, with $k \neq j$.

If we call $\psi_j(T)$ the solution (5.92) at time T, with $f = u_{oj}$, and put

$$c_{jk} =< u_{ok}, \psi_j(T) > \, , \tag{5.93}$$

we know that the sought transition probability is

$$P_{jk} = |c_{jk}|^2 = | < u_{ok}, \psi_j(T) > |^2 \, . \tag{5.94}$$

We first notice that, if $f = u_{oj}$, $\psi_{oj}(t)$ is always proportional to it

$$\psi_{oj}(t) = e^{-i\frac{H_0 t}{\hbar}} u_{oj} = e^{-i\frac{E_{oj} t}{\hbar}} u_{oj} \ , \tag{5.95}$$

so that, recalling that $\psi(T) = \psi_{oj}(T) + \varepsilon\delta\psi_j(T)$ and that

$$< u_{ok}, \psi_{oj}(T) >= e^{-i\frac{E_{oj} t}{\hbar}} < u_{ok}, u_{oj} >= 0$$

we get

$$c_{jk} = \varepsilon < u_{ok}, \delta\psi_j(T) > \ . \tag{5.96}$$

Moreover, in the present case

$$\delta\psi_j(T) = -\frac{i}{\hbar}\int_0^T e^{-i\frac{H_0(T-\tau)}{\hbar}} \delta H(t) e^{-i\frac{E_{oj}\tau}{\hbar}} u_{oj} d\tau \ ,$$

so that

$$c_{jk} = -\frac{i\varepsilon}{\hbar}\int_0^T < u_{ok}, e^{-i\frac{H_0(T-\tau)}{\hbar}} \delta H(\tau) u_{oj} > e^{-i\frac{E_{oj}\tau}{\hbar}} d\tau \ . \tag{5.97}$$

On the other hand, recalling that the L^2 product is antilinear in the first element, we get

$$\begin{aligned} < u_{ok}, e^{-i\frac{H_0(T-\tau)}{\hbar}} \delta H(T) u_{oj} > &= < e^{i\frac{H_0(T-\tau)}{\hbar}} u_{ok}, \delta H(\tau) u_{oj} >= \\ &= e^{-i\frac{E_{ok}(T-\tau)}{\hbar}} < u_{ok}, \delta H(\tau) u_{oj} > \ . \end{aligned}$$

Finally we obtain

$$c_{jk} = -i\frac{\varepsilon e^{-i\frac{E_{ok}T}{\hbar}}}{\hbar}\int_0^T e^{-i\frac{(E_{oj}-E_{ok})\tau}{\hbar}} < u_{ok}, \delta H(\tau) u_{oj} > d\tau \ , \tag{5.98}$$

or

$$P_{jk} = \frac{\varepsilon^2}{\hbar^2}\left|\int_0^T e^{-i\frac{(E_{oj}-E_{ok})\tau}{\hbar}} < u_{ok}, \delta H(\tau) u_{oj} > d\tau\right|^2 \ . \tag{5.99}$$

One consequence of (5.99) is that, since the conjugate of a complex number has the same modulus as the number itself, we have that transition probabilities are symmetric, namely

$$P_{jk} = P_{kj}. \tag{5.100}$$

Another consequence of (5.99) is that if we assume $\delta H(t)$ to be generated by a radiation with a frequency ω_{jk} in resonance with the jump $j \to k$, i.e.

$$\omega_{jk} = \frac{E_{ok} - E_{oj}}{\hbar}, \tag{5.101}$$

then we can hypothesize that

$$\delta H(t) = e^{i\omega t}\delta H_0 \tag{5.102}$$

and substitute in (5.99). Putting

$$\Omega_{jk} = \frac{\varepsilon < u_{ok}, \delta H_0 u_{oj} >}{\hbar},$$

we then derive

$$P_{jk} = |\Omega_{jk}|^2 T^2 . \tag{5.103}$$

As we can see, (5.103) clearly displays the perturbation character of this calculation because for a long T this expression will give numbers larger than 1. However in the next chapter we will see, after a different calculation, that (5.103) can be considered correct for small T.

References

1. Caldirola P, Cirelli R, Prosperi GM (1982) Introduzione alla fisica teorica. UTET, Torino
2. Messiah A (1965) Mécanique quantique 1–2. Dunod, Paris
3. Pauli W (1962) Meccanica ondulatoria. Boringhieri, Torino
4. Nikiforov AF, Uvarov VB (1975) On a new approach to the theory of special functions. Mat Sb (N.S.) 98(140), 4(12), 578–590
5. Nikiforov AF, Uvarov VB (1988) Special functions of mathematical physics. Birkhauser Basel https://doi.org/10.1007/978-1-4757-1595-8
6. Caldirola P (1966) Istituzioni di fisica, teorica edn. Viscontea, Milano
7. Dirac PAM (1959) I principi della meccanica quantistica. Boringhieri, Torino
8. Kato T (1966) Perturbation theory of linear operators. Springer, Principles of Mathematical Sciences
9. Kato T (1982) A short introduction to the perturbation theory of linear operators. Springer

Chapter 6
The Quantum Measurement of Gravity

6.1 Introduction

There are several methods that use the quantum interaction between laser beams and atoms to measure gravity; here we shall refer to a method based on the theory of Ramsey on stimulated level transitions, already developed in the 50s, that, however, was first used in 1992 by Casevich and Chu, demonstrating the feasibility of gravity measurements. Of course, there are numerous smaller effects and technical particulars that will enter into this description. Yet, as we claimed several times, here the target is to give a motivated picture of the ideas that are behind this approach.

We will first analyse an atom in a stationary e.m. field at optical frequency (laser beam) showing that in this case the interaction field—atom can be described by a Hamiltonian term called *dipole coupling*.

Then we will analyse the time evolution of the dynamic system of the atom by constraining it to a Hilbert subspace containing only the two (or three) unperturbed eigenstates, that are interacting with a quasi-resonant radiation. This results in the so-called *Rabi equations* that have the merit to describe, though approximately, the evolution of the system on any time span. Based on such equations, the theoretical possibility of atom interferometry is established.

Next we will show how the above theory can be applied to a sample of atoms launched in a ballistic trajectory, bombarded by optical pulses, that through the interaction between internal and external degrees of freedom gives the measurement of gravity as a byproduct.

In a final section, a number of important technical improvements on the previous theory will be described. These improvements, as a matter of fact, allow to reach the very high accuracy that has already been demonstrated in the year 2000.

The matter of this chapter, representing the final goal of the book, can be found spread in many articles in Physics literature. Yet the material in the book on Atom Interferometry [1] collecting contributions from different Authors, has been fundamental for our scope.

F. Sansò and F. Migliaccio, *Quantum Measurement of Gravity for Geodesists and Geophysicists*, Springer Geophysics,
https://doi.org/10.1007/978-3-030-42838-9_6

6.2 The Dipole Approximation of the Optical Radiation—Atom Interaction

Let us consider an electron of an atom, which is in a stationary state, related to one of the external orbits that we will indicate with an index g to mean ground state. The internal orbits are occupied by other electrons in stationary states and a "selection rule", which we have not discussed in these notes, forbids our electron to jump into one of these states. On the contrary, other higher energy levels, and the corresponding states, are open to the electron if there is a supply of energy from outside, supporting this transition. We will focus our attention on one of these levels in particular, which we will denote with the index e, corresponding to an excited state.

If at a certain moment in time we switch an e.m. signal on, the electron, in fact the atom as a whole, has one interaction with the e.m. field that is described by the Hamiltonian (4.91), to which we add the coulomb interaction with the nucleus V, namely

$$H = \frac{1}{2m}\left(-\hbar^2 \Delta + i\hbar\frac{e}{c}(\nabla \cdot \boldsymbol{A}) + 2i\hbar\frac{e}{c}\boldsymbol{A} \cdot \nabla + \frac{e^2}{c^2}A^2\right) - \varphi + V. \quad (6.1)$$

The field $\boldsymbol{A}$ is the vector potential and φ the scalar potential of the e.m. field. Since we are in empty space, i.e. we neglect the field generated by the electron with respect to the much stronger external field, we can use a gauge in which (see Remark 2.5)

$$\varphi = 0\,,\ \ \nabla \cdot \boldsymbol{A} = 0\,, \quad (6.2)$$

so that the electric field and the magnetic field will be given by

$$\boldsymbol{E} = -\frac{1}{c}\frac{\partial A}{\partial t}\,,\ \ \boldsymbol{H} = \nabla \wedge \boldsymbol{A}\,. \quad (6.3)$$

Returning to (6.1) we see that the terms $\nabla \cdot \boldsymbol{A}$ and φ can be cancelled. Moreover, the term $\frac{e^2}{c^2}A^2$ is known to be much smaller than the others and we can neglect it. Therefore, our Hamiltonian can be put in the form

$$H = H_0 + \delta H, \quad (6.4)$$

where H_0 expresses the interaction of the electron with the atom to which it is bound in a stationary state g, while

$$\delta H = 2i\hbar\frac{e}{2mc}\boldsymbol{A} \cdot \nabla = \frac{i\hbar e}{mc}\boldsymbol{A} \cdot \nabla \quad (6.5)$$

expresses the interaction of the electron with the e.m. field.

It is our purpose to prove that, when applied to eigenfunctions of H_0, that go rapidly to zero beyond the radius of the atom, and when $\boldsymbol{A}$ refers to a plane monochromatic wave at optical frequency, we can apply the approximation

$$\delta H \cong -\boldsymbol{E} \cdot (e\boldsymbol{r}) = -\boldsymbol{E} \cdot \boldsymbol{d}, \tag{6.6}$$

where $\boldsymbol{r}$ is the radius vector of the position of the electron in a reference system centred at the nucleus of the atom and $\boldsymbol{d} = e\boldsymbol{r}$ its electric dipole moment.

More precisely, let us formulate the following proposition.

Proposition 6.1 *Let* $\{u_{oi}\}(i = e, g)$ *denote the unperturbed eigenstates of an optical (external) electron of an atom and let* δH*, given by (6.5), be a perturbing Hamiltonian due to an e.m. field in the optical frequency range (i.e. wavelengths between 0,1 and 1 μm), then we have, with a high approximation level,*

$$< u_{oi}, \delta H u_{oj} > \cong < u_{oi}, (-\boldsymbol{E} \cdot \boldsymbol{d}) u_{oj} > , \tag{6.7}$$

proving that in such a set up formula (6.6) holds.

Proof For a plane wave we can write

$$\boldsymbol{A} = \boldsymbol{A}_0 Re[e^{i(\boldsymbol{k}\cdot\boldsymbol{r} - \omega t - \alpha)}], \tag{6.8}$$

where Re denotes the real part of a complex number, $\omega = c|\boldsymbol{k}|$ and α is an arbitrary constant phase factor. By defining

$$2\boldsymbol{U}_0 = \boldsymbol{A}_0 e^{-i\alpha}$$

we can write (6.8) as

$$\begin{aligned} \boldsymbol{A} &= 2Re[\boldsymbol{U}_0 e^{i(\boldsymbol{k}\cdot\boldsymbol{r} - \omega t)}] = \\ &= [\boldsymbol{U}_0 e^{i(\boldsymbol{k}\cdot\boldsymbol{r} - \omega t)} + \boldsymbol{U}_0^* e^{-i(\boldsymbol{k}\cdot\boldsymbol{r} - \omega t)}] . \end{aligned} \tag{6.9}$$

Since in (6.9) we see two components, one with frequency $\frac{\omega}{2\pi}$, the other with frequency $-\frac{\omega}{2\pi}$, and since we will study our system with a frequency almost resonant with the internal atomic frequency (recall (3.16))

$$\omega_{ge} = \frac{E_{oe} - E_{og}}{\hbar} > 0 \tag{6.10}$$

that corresponds to a jump between the two unperturbed energy levels E_{og}, E_{oe}, we can neglect the conjugate term in (6.9), that does not take part into energy exchanges, and directly write

$$\boldsymbol{A} = \boldsymbol{U}_0 e^{i(\boldsymbol{k}\cdot\boldsymbol{r} - \omega t)} \tag{6.11}$$

into (6.5).

Moreover, let us consider the integral

$$I_{ij} = \int u_{oi}^* \boldsymbol{A} \cdot \nabla u_{oj} dx = \\ = \boldsymbol{U}_0 e^{-i\omega t} \cdot \int u_{oi}^* (\nabla u_{oj}) e^{i\boldsymbol{k}\cdot\boldsymbol{r}} dx \; ; \tag{6.12}$$

since the eigenstates u_{oi}, u_{oj} decrease exponentially beyond the radius R_0 of the atom, that is of the order of 10^{-7} mm, while $(\boldsymbol{k}) = \frac{2\pi}{\lambda}$ is of the order of 10^3 mm^{-1} we see that

$$O(\boldsymbol{k} \cdot \boldsymbol{r}) < 10^{-4} \tag{6.13}$$

in the region where the integrand of (6.12) is significantly non-zero. Therefore, we can safely put in our atomic region

$$e^{i\boldsymbol{k}\cdot\boldsymbol{r}} \cong 1 \; , \; \boldsymbol{A} = \boldsymbol{U}_0 e^{-i\omega t} \tag{6.14}$$

and write (6.12) as

$$\begin{cases} I_{ij} = \boldsymbol{U}_0 e^{-i\omega t} \cdot \boldsymbol{J}_{ij} \\ \boldsymbol{J}_{ij} = \int u_{oi}^* \nabla u_{oj} dw \; . \end{cases} \tag{6.15}$$

Now we use the identity

$$\boldsymbol{J}_{ij} = -\frac{1}{2} \int u_{oi}^* [\boldsymbol{r} \Delta u_{oj} - \Delta(\boldsymbol{r} u_{oj})] dx \; . \tag{6.16}$$

Such a relation holds because (recall $\Delta \boldsymbol{r} = 0, \; \nabla \boldsymbol{r} = I$)

$$\Delta(\boldsymbol{r} u_{oj}) = (\Delta \boldsymbol{r}) u_{oj} + 2\nabla \boldsymbol{r} \cdot \nabla u_{oj} + \boldsymbol{r} \Delta u_{oj} = \\ = 2\nabla u_{oj} + \boldsymbol{r} \Delta u_{oj}$$

that substituted into (6.16) gives (6.15).

On the other hand

$$\int u_{oi}^* \Delta(\boldsymbol{r} u_{oj}) dx \equiv \int (\Delta u_{oi}^*) \boldsymbol{r} u_{oj} dx$$

because out of the boundary of the atom u_{oi}, u_{oj} are practically zero. So (6.16) becomes

$$\boldsymbol{J}_{ij} = -\frac{1}{2} \int \boldsymbol{r} [u_{oi}^* \Delta u_{oj} - (\Delta u_{oi}^*) u_{oj}] dx \; . \tag{6.17}$$

If we write the equations for the stationary states u_{oi}, u_{oj} in the form

$$\begin{cases} \Delta u_{oj} = -\frac{2m}{\hbar^2} E_{oj} u_{oj} + \frac{2m}{\hbar^2} V u_{oj} \\ \Delta u^*_{oi} = -\frac{2m}{\hbar^2} E_{oi} u^*_{oi} + \frac{2m}{\hbar^2} V u^*_{oi} \,, \end{cases}$$

and we multiply the first by $\boldsymbol{r} u^*_{oi}$, the second by $\boldsymbol{r} u_{oj}$ and subtract, returning to (6.17), we find

$$\begin{aligned} \boldsymbol{J}_{ij} &= -\frac{1}{2} \int \frac{(E_{oi} - E_{oj})2m}{\hbar^2} \boldsymbol{r} u^*_{oi} u_{oj} dx = \\ &= -\frac{m}{\hbar} \omega_{ji} \int \boldsymbol{r} u^*_{oi} u_{oj} dx \end{aligned} \tag{6.18}$$

with $\omega_{ji} = \hbar^{-1}(E_{oi} - E_{oj})$.

So, recalling (6.15), we get

$$I_{ij} = -\frac{m}{\hbar} \omega_{ji} \boldsymbol{U}_0 e^{-i\omega t} \cdot \int \boldsymbol{r} u^*_{oi} u_{oj} dx \tag{6.19}$$

and finally

$$\begin{aligned} < u_{oi}, \delta H u_{oj} > &= i \frac{\hbar e}{mc} \boldsymbol{U}_0 e^{-i\omega t} \cdot \int u^*_{oi} \nabla u_{oj} dx = \\ &= -i \frac{\hbar e}{mc} e^{-i\omega t} \boldsymbol{U}_0 \cdot \frac{m}{\hbar} \omega_{ji} \int \boldsymbol{r} u^*_{oi} u_{oj} dx = \\ &= -i \frac{\omega_{ji}}{c} e^{-i\omega t} \boldsymbol{U}_0 \cdot \int (e\boldsymbol{r}) u^*_{oi} u_{oj} dx \,. \end{aligned} \tag{6.20}$$

Now, recalling (6.3) and (6.14), we see that, in this context, i.e. in the volume of the atom,

$$\boldsymbol{E} = -\frac{1}{c} \frac{\partial \boldsymbol{A}}{\partial t} = i \frac{\omega}{c} e^{-i\omega t} \boldsymbol{U}_0$$

so that (6.20) can be written as

$$< u_{oi}, \delta H u_{oj} >= -\frac{\omega_{ji}}{\omega} \boldsymbol{E} \cdot \int (e\boldsymbol{r}) u^*_{oi} u_{oj} dx \,. \tag{6.21}$$

If we put $i = j$ in (6.21), since $\omega_{ii} = 0$, we find

$$< u_{oi}, \delta H u_{oi} >= 0 \,.$$

On the other hand, considering that $\boldsymbol{E}$ is practically constant in the region of the atom, we have

$$\int \boldsymbol{E} \cdot (e\boldsymbol{r}) u^*_{oi} u_{oi} d\boldsymbol{x} = e\boldsymbol{E} \cdot \int \boldsymbol{r} |u_{oi}|^2 dx = e\boldsymbol{E} \cdot \boldsymbol{r}_b,$$

where $\boldsymbol{r}_b$ is just the position of the geometric barycentre of the state i; but this is zero as a consequence of our choice of coordinate system, i.e.

$$\int \boldsymbol{E} \cdot (e\boldsymbol{r})|u_{oi}^*|^2 dx = 0$$

and we can say that in this case

$$< u_{oi}, \delta H u_{oi} >=< u_{oi}, -\boldsymbol{E} \cdot \boldsymbol{d}\, u_{oi} >= 0 \,, \tag{6.22}$$

namely (6.7) is valid in this particular case.

If, on the contrary, $i = e$ and $j = g$, one has

$$\omega_{ji} = \omega_{ge} \cong \omega$$

because of the almost resonance hypothesis done.

So from (6.21) we see that

$$< u_{oe}, \delta H u_{og} >\cong< u_{oe}, (-\boldsymbol{E} \cdot \boldsymbol{d}) u_{og} > \ . \tag{6.23}$$

The relations (6.22), (6.23) complete the proof of (6.7). □

Remark 6.1 There is a much simpler proof of (6.7) which, however, is based more heavily on the classical analysis of e.m. forces acting on the electron, and as such is not giving a perfect control of the approximations assumed.

In fact, if we consider the electric field $\boldsymbol{E}$ depending only on t and not on $\boldsymbol{x}$ in the volume of the atom, we can directly compute the potential generating $\boldsymbol{E}$, namely

$$V(\boldsymbol{x}) = -\boldsymbol{E} \cdot \boldsymbol{x} \,, \tag{6.24}$$

in fact, since $\boldsymbol{E}$ does not depend on $\boldsymbol{x}$, it is clearly

$$-\nabla V = \boldsymbol{E} \,. \tag{6.25}$$

But then the perturbation of the energy of the electron can be written as

$$\delta H = eV(\boldsymbol{x}) = -\boldsymbol{E} \cdot e\,\boldsymbol{x} = -\boldsymbol{E} \cdot \boldsymbol{d} \,,$$

which is the same as (6.23).

Let us further notice that in this picture we simply ignore any magnetic interaction, which by the way is small, but in the general treatment presented before, the interaction with the magnetic field is automatically accounted for by the use of the vector potential $\boldsymbol{A}$.

6.3 Rabi Equation

The Rabi approach to the solution of Schrödinger equation for an atom interacting with an optical plane wave is essentially to constrain the description of the dynamic system to possess only two degrees of freedom, corresponding to the energies E_g, E_e because all the other degrees of freedom (i.e. eigenstates making up the Hilbert subspace corresponding to all the bound states) do not participate in the interaction (at least approximately). In fact, the pulsation (we will call it in short the *frequency*) ω of the e.m. wave is close to resonance to the transition $g \leftrightarrow e$, namely putting

$$\omega = \omega_{ge} + \delta = \frac{E_e - E_g}{\hbar} + \delta \tag{6.26}$$

the quantity δ, called the *detuning*, has to satisfy

$$|\delta| \ll |\omega_e| \ , \ |\omega_g| \ , \tag{6.27}$$

so that if the wave is formed by a packet centered in ω, it can only exchange energy with the two states g, e because of basic quantum conditions (cf. Sect. 3.4) and cannot interact with other internal energy levels. So, if we call u_g, u_e the free eigenstates corresponding to E_g, E_e, respectively, we find that, for the dynamics of our interest, the dynamic system is described by the Hamiltonian

$$H\psi = E_g u_g < u_g, \psi > + E_e u_e < u_e, \psi > + (-\boldsymbol{E} \cdot \boldsymbol{d})\psi = H_0\psi + \delta H\psi \ . \tag{6.28}$$

Therefore, the time evolution of the state $\psi(t)$ is controlled by Schrödinger equation

$$i\hbar\dot{\psi} = (H_0 + \delta H)\psi \ . \tag{6.29}$$

Indeed, even if $\psi_0 = \psi(0)$ is in the space Span $\{u_g, u_e\}$, strictly speaking $\delta H \psi(t)$ needs not to be in the same space. However, as claimed before, the probability that δH induces a transition to an external eigenstate is so small that we can safely put

$$\psi(t) = a_g(t)u_g + a_e(t)u_e \ . \tag{6.30}$$

Therefore, (6.29) becomes

$$\begin{aligned} i\hbar(\dot{a}_g u_g + \dot{a}_e u_e) &= (E_g a_g u_g + E_e a_e u_e) + \\ &+ (a_g \delta H u_g + a_e \delta H u_e) \ . \end{aligned} \tag{6.31}$$

Let us take the scalar product of (6.31) with u_e and with u_g, respectively, and recall that (see (6.22))

$$< u_g, \delta H u_g >= 0 \ , \ < u_e, \delta H u_e >= 0 \ ;$$

the result is

$$\begin{cases} i\hbar\dot{a}_e = E_e a_e + V_{ge} a_g \\ i\hbar\dot{a}_g = E_g a_g + V_{ge}^* a_e \end{cases} \tag{6.32}$$

$$(V_{ge} =< u_e, (-\boldsymbol{E} \cdot \boldsymbol{d}) u_g >= \cos(\omega t + a) < u_e, -e\boldsymbol{E}_0 \cdot \boldsymbol{r} u_g >)$$

To (6.32) we associate the initial value problem

$$a_e(0) =< u_e, \psi_0 > \ , \ a_g(0) =< u_g, \psi_0 > \ ; \tag{6.33}$$

indeed the coefficients a_e, a_g have to satisfy at any time t the unitarity condition

$$|a_e|^2 + |a_g|^2 = 1 \tag{6.34}$$

that corresponds to the condition $\| \psi(t) \|^2 = 1$. It is convenient to eliminate $\hbar$ from (6.32) by setting

$$\hbar\omega_e = E_e \ , \ \hbar\omega_g = E_g \ , \ \hbar\cos(\omega t + \varphi)\Omega_{ge} = V_{ge} \ . \tag{6.35}$$

Naturally the quantities $\omega_e, \omega_g, \Omega_{ge}$ are dimensionally frequencies. To simplify the notation we will simply denote Ω_{ge} by Ω. We note as well that while ω_e, ω_g are real, Ω is in general complex as V_{ge}; as a matter of fact it will be

$$\hbar\Omega =< u_e, (-e\boldsymbol{E}_0 \cdot \boldsymbol{r}) u_g > \ . \tag{6.36}$$

Moreover, since V_{ge} expresses a perturbation of the energy levels, we shall assume that the inequalities hold

$$|\Omega| \ll \omega_e, \omega_g \ . \tag{6.37}$$

Given the above positions and remarks, we write (6.32) in the form

$$\begin{cases} i\dot{a}_e = \omega_e a_e + \cos(\omega t + \varphi)\Omega a_g \\ i\dot{a}_g = \omega_g a_g + \cos(\omega t + \varphi)\Omega^* a_e . \end{cases} \tag{6.38}$$

We first of all factor the rapidly oscillating part of the solution (a_g, a_e) of (6.38), by putting

$$\begin{cases} a_e(t) = e^{-i\omega_e t} b_e(t) \\ a_g(t) = e^{-i\omega_g t} b_g(t) . \end{cases} \tag{6.39}$$

Substituting in (6.38) and simplifying, we get

$$\begin{cases} i\dot{b}_e(t) = \Omega \cos(\omega t + \varphi) e^{i\omega_{ge}t} b_g(t) \\ i\dot{b}_g(t) = \Omega^* \cos(\omega t + \varphi) e^{-i\omega_{ge}t} b_e(t). \end{cases} \tag{6.40}$$

If we write the first of (6.40) as

$$ib_e(t) = ib_e(0) + \Omega \int_0^t \frac{e^{i(\omega t'+\varphi)} + e^{-i(\omega t'+\varphi)}}{2} e^{i\omega_{ge}t'} b_g(t')dt'$$

we realize that $b_g(t')$ is multiplied once by an exponential with pulsation $-\omega + \omega_{ge} = -\delta$, namely with a low frequency, and another time by the high-frequency pulsation $\omega + \omega_{ge} = 2\omega_{ge} + \delta$. Therefore, only the second term multiplied by $b_g(t')$ will give a significant integral, while the first one will be much smaller and we will neglect it. A similar reasoning holds for the second of (6.40), so that we can approximate it as

$$\begin{cases} i\dot{b}_e = \frac{1}{2}\Omega e^{-i(\delta t+\varphi)} b_g = \frac{1}{2}\Gamma e^{-i(\delta t+\gamma)} b_g \\ i\dot{b}_g = \frac{1}{2}\Omega^* e^{i(\delta t+\varphi)} b_e = \frac{1}{2}\Gamma e^{i(\delta t+\gamma)} b_e \,, \end{cases} \tag{6.41}$$
$$(\delta = \omega - \omega_{ge} = \omega - (\omega_e - \omega_g))$$

where we have put

$$\Omega = \Gamma e^{i\alpha} \,,\ (\Gamma = |\Omega|) \,;\ \gamma = \varphi - \alpha \,. \tag{6.42}$$

The final transformation

$$\begin{cases} b_e = e^{-i\frac{1}{2}(\delta t+\gamma)} c_e \\ b_g = e^{i\frac{1}{2}(\delta t+\gamma)} c_g \end{cases} \tag{6.43}$$

brings (6.41) into

$$i\dot{c} = i \begin{vmatrix} \dot{c}_e \\ \dot{c}_g \end{vmatrix} = \frac{1}{2} \begin{vmatrix} -\delta & \Gamma \\ \Gamma & \delta \end{vmatrix} c = \frac{1}{2} D\, c \,. \tag{6.44}$$

This is a linear system with constant coefficients of which we know the solution, i.e.,

$$c(t) = e^{-i\frac{1}{2}Dt} c(0) \,. \tag{6.45}$$

In order to practically compute (6.45) we need the spectral decomposition of D, namely

$$D = \lambda_1 \xi_1 \xi_1^\dagger + \lambda_2 \xi_2 \xi_2^\dagger$$

with $\lambda_{1,2}, \xi_{1,2}$ eigenvalues and eigenvectors of D. It is convenient to put

$$\Omega_r = \sqrt{\Gamma^2 + \delta^2}\ , \ \sin\vartheta = \frac{\Gamma}{\Omega_r}\ , \ \cos\vartheta = -\frac{\delta}{\Omega_r}\ ; \tag{6.46}$$

in particular Ω_r is known as *off-resonant Rabi frequency*. In terms of the above quantities we find

$$\lambda_1 = \Omega_r\ , \ \boldsymbol{\xi}_1 = \begin{vmatrix} \cos\frac{\vartheta}{2} \\ \sin\frac{\vartheta}{2} \end{vmatrix}\ ; \ \lambda_2 = -\Omega_r\ , \ \boldsymbol{\xi}_2 = \begin{vmatrix} \sin\frac{\vartheta}{2} \\ -\cos\frac{\vartheta}{2} \end{vmatrix}\ . \tag{6.47}$$

so that we can write

$$\begin{aligned} \boldsymbol{c}(t) = e^{-i\frac{\Omega_r t}{2}} \left[\cos\frac{\vartheta}{2} c_e(0) + \sin\frac{\vartheta}{2} c_g(0)\right] \begin{vmatrix} \cos\frac{\vartheta}{2} \\ \sin\frac{\vartheta}{2} \end{vmatrix} + \\ + e^{i\frac{\Omega_r t}{2}} \left[\sin\frac{\vartheta}{2} c_e(0) - \cos\frac{\vartheta}{2} c_g(0)\right] \begin{vmatrix} \sin\frac{\vartheta}{2} \\ -\cos\frac{\vartheta}{2} \end{vmatrix}\ . \end{aligned} \tag{6.48}$$

Let us note that

$$\begin{cases} c_e(0) = e^{i\frac{\gamma}{2}} b_e(0) = e^{i\frac{\gamma}{2}} a_e(0) \\ c_g(0) = e^{-i\frac{\gamma}{2}} b_g(0) = e^{-i\frac{\gamma}{2}} a_g(0), \end{cases} \tag{6.49}$$

which is important because $a_e(0)$, $a_g(0)$ are given data as we must assign the initial state ψ_0 and $a_{e,g}(0) =< u_{e,g}, \psi_0 >$.

By developing the products of (6.48) one gets

$$\begin{cases} c_e(t) = \left(\cos\frac{\Omega_r t}{2} - i\cos\vartheta\sin\frac{\Omega_r t}{2}\right) c_e(0) - i\sin\frac{\Omega_r t}{2}\sin\vartheta\, c_g(0) \\ c_g(t) = -i\sin\frac{\Omega_r t}{2}\sin\vartheta\, c_e(0) + \left(\cos\frac{\Omega_r t}{2} + i\cos\vartheta\sin\frac{\Omega_r t}{2}\right) c_g(0)\ . \end{cases} \tag{6.50}$$

By using (6.43), (6.49) and (6.50), we reconstruct the dynamics in terms of the "slow" coefficients

$$\begin{cases} b_e(t) = e^{-i\frac{\delta t}{2}} \left\{\left(\cos\frac{\Omega_r t}{2} - i\cos\vartheta\sin\frac{\Omega_r t}{2}\right) b_e(0) - ie^{-i\gamma}\sin\frac{\Omega_r t}{2}\sin\vartheta\, b_g(0)\right\} \\ b_g(t) = e^{i\frac{\delta t}{2}} \left\{-ie^{i\gamma}\sin\vartheta\sin\frac{\Omega_r t}{2} b_e(0) + \left(\cos\frac{\Omega_r t}{2} + i\cos\vartheta\sin\frac{\Omega_r t}{2}\right) b_g(0)\right\}. \end{cases} \tag{6.51}$$

Finally, reproducing all our steps (6.39), (6.43) we can reconstruct the time evolution of $a_e(t)$, $a_g(t)$ that results

$$\begin{cases} a_e(t) = e^{-i(\omega_e + \frac{\delta}{2})t} \left\{\left(\cos\frac{\Omega_r t}{2} - i\cos\vartheta\sin\frac{\Omega_r t}{2}\right) a_e(0) + \right. \\ \qquad\qquad \left. -ie^{-i\gamma}\sin\vartheta\sin\frac{\Omega_r t}{2} a_g(0)\right\} \\ a_g(t) = e^{-i(\omega_g - \frac{\delta}{2})t} \left\{-ie^{i\gamma}\sin\frac{\Omega_r t}{2} a_e(0) + \right. \\ \qquad\qquad \left. \left(\cos\frac{\Omega_r t}{2} + i\cos\vartheta\sin\frac{\Omega_r t}{2}\right) a_g(0)\right\}\ . \end{cases} \tag{6.52}$$

At this point it is nice to see that if the light is perfectly resonant, i.e. $\delta = 0$, $\Omega_r = |\Omega|$, $\sin\vartheta = 1$, and we start at $t = 0$ in the g state, $Q_e(0) = 0$, $a_g(0) = 1$, when we compute $|a_e(t)|^2$, namely the probability that the system at time t has transited to the state e, we get

$$|a_e(t)|^2 = \left| -ie^{-i\gamma} \sin \frac{|\Omega| t}{2} \right|^2 = \frac{1}{2}(1 - \cos |\Omega| t) \; ; \tag{6.53}$$

as we see, the state of the system swaps from g to e and vice versa with frequency Ω; this process is called the *Rabi oscillation*.

It is also interesting to observe that (6.52), taken for short times (small $|\Omega|(t)$), gives

$$|a_e(t)|^2 \cong \frac{1}{4} |\Omega|^2 t^2$$

which is essentially equal to (5.103). The factor $\frac{1}{4}$ is just due to the fact that here we have used the real-valued perturbation (see δH in (6.28)), where the amplitude of the field $\boldsymbol{E}_0$ is split 1/2 for the frequency ω, 1/2 for the frequency $-\omega$, and the corresponding intensities are $\frac{1}{4}$, $\frac{1}{4}$.

In the next two examples we will see how, exploiting (6.50), (6.52), we can use a light pulse to act on the atom ψ-wave as a splitter or as a mirror, depending on the length of the pulse. In this way, we will show that light pulses manipulation can give the same effects as those of a crystal interferometer for light. These will be the main tools for atom interferometry.

Remark 6.2 For the sake of simplicity, it is useful to remark that, since by (6.39) we have

$$|a_e(t)|^2 = |b_e(t)|^2 \; , \; |a_g(t)|^2 = |b_g(t)|^2$$

as well as

$$\begin{cases} a_e(0) = b_e(0) \\ a_g(0) = b_g(0) \; ; \end{cases}$$

we can continue the description of the dynamic behaviour of our system only in terms of the "slowly oscillating" coefficients $b_e(t), b_g(t)$.

Example 6.1 Assume that an atom starting at $t = 0$ in the state g is illuminated by a light pulse of duration τ; after τ the light is switched off and we would like to know what is the physical state ψ of the atom when we suppose that

$$\Omega_r \tau = \frac{\pi}{2} \; . \tag{6.54}$$

In other words, we would like to know $|b_e(\tau)|^2$, $|b_g(\tau)|^2$ when $b_e(0) = 0$, $b_g(0) = 1$ and τ satisfies (6.54). In this case we find

$$\begin{cases} b_e(\tau) = e^{-i\frac{\delta}{2}\tau}\left[-ie^{-i\gamma}\sin\vartheta\frac{1}{\sqrt{2}}\right] \\ b_g(\tau) = e^{i\frac{\delta}{2}\tau}\left[\frac{1}{\sqrt{2}} + i\cos\vartheta\frac{1}{\sqrt{2}}\right] . \end{cases} \tag{6.55}$$

Therefore, if the pulse is exactly resonant, $\delta = 0$, $\sin\vartheta = 1$, $\cos\vartheta = 0$, we notably get

$$|b_e(\tau)|^2 = |b_g(\tau)|^2 = \frac{1}{2} ,$$

namely the state of the system is split with probability 1/2 between the two stationary states e and g. So the pulse of duration $\tau = \frac{\pi}{2}\Omega_r^{-1}$ acts as a material splitter, e.g. like a crystal lattice.

We can notice as well that if, on the contrary, we start from the e level, namely $b_e(0) = 1$, $b_g(0) = 0$, from (6.51) we get a relation reciprocal to (6.54), namely

$$\begin{cases} b_e(\tau) = e^{-i\frac{\delta}{2}\tau}\left[\frac{1}{\sqrt{2}} - i\cos\vartheta\frac{1}{\sqrt{2}}\right] \\ b_g(\tau) = e^{i\frac{\delta}{2}\tau}\left[-ie^{i\gamma}\sin\vartheta\frac{1}{\sqrt{2}}\right] . \end{cases} \tag{6.56}$$

In particular, when we have perfect resonance, i.e. $\sin\vartheta = 1$, $\cos\vartheta = 0$, we get again

$$|b_e(\tau)|^2 = |b_g(\tau)|^2 = \frac{1}{2}$$

meaning that we have probability 1/2 to remain in e and probability 1/2 to fall on g. In this case, of course, a photon is emitted by the atom that decreases its energy level; this is called *stimulated Raman transition* in literature. This symmetry confirms also our findings of Sect. 5.7 (see (5.100)).

Example 6.2 We repeat the same calculation as in Example 6.1, but this time we take a pulse with double duration 2τ, i.e. such that

$$\Omega_r(2\tau) = \pi . \tag{6.57}$$

In this case indeed $\cos\frac{\Omega_r 2\tau}{2} = 0$, $\sin\frac{\Omega_r 2\tau}{2} = 1$ and assuming again $\delta \cong 0$, $\sin\vartheta \cong 1$, $\cos\vartheta \cong 0$ we find

$$\begin{cases} b_e(2\tau) = e^{-i(\frac{\delta}{2})2\tau}\left[-ie^{-i\gamma}a_g(0)\right] \\ b_g(2\tau) = e^{i\frac{\delta}{2}\cdot 2\tau}\left[-ie^{i\gamma}a_e(0)\right] . \end{cases} \tag{6.58}$$

But this implies

$$|b_e(2\tau)|^2 = |b_g(0)|^2,$$
$$|b_g(2\tau)|^2 = |b_e(0)|^2 ,$$

namely if you start from g you end up with e and vice versa. Therefore, the function of a material mirror is implemented.

Remark 6.3 Let us observe that if we switch off the e.m. field, Eqs. (6.50), (6.51), with $\delta = 0$, $\Omega_r = 0$, simply become

$$\begin{cases} a_e(t) = e^{-i\omega_e t} a_e(0) \\ a_g(t) = e^{-i\omega_g t} a_g(0) \ , \end{cases}$$

that in terms of the slow coefficients $b_e(t), b_g(t)$, are simply

$$\begin{cases} b_e(t) = b_e(0) \\ b_g(t) = b_e(0) \ , \end{cases} \tag{6.59}$$

i.e. when the field is switched off, the “slow” coefficients become constant.

6.4 Adding Momentum Exchange to the Atom–Photon Interaction

In the previous section the description of the interaction e.m. field—atom has been developed for an atom, so to say, still in space, a fact that has been translated by placing the origin at the nucleus of the atom. In this way, however, we completely lose the effects of a fundamental law of physics, namely the conservation of momentum. In fact, we know that if an electromagnetic quantum of wavelength λ is absorbed or emitted by the atom, its momentum, that according to (3.30) has a modulus given by

$$p = \frac{h}{\lambda} \ ,$$

has to be transmitted to the atom, so that, if the atom is at rest at $t < 0$, and we have the transition at $t = 0$, the idea that it continues to be still for $t > 0$ is untenable. Naturally, this was not a mistake, rather the previous reasoning, developed by Ramsey as early as 1950, was framed in the context of the interaction of an atom with a microwave e.m. radiation. In that case the momentum exchange was negligible, but since in this context we analyse the interaction at optical frequency, the modulus of the momentum of a radiation quantum has to be multiplied by a factor $10^3 \div 10^4$ and the conservation of momentum has to be accounted for.

We will analyse the change in the solution of the equations for the “slow” coefficients b_g, b_e (i.e. (6.41), (6.42)) taking into account the degrees of freedom that describe the kinematic behaviour of the barycentre of the atom. We will concentrate first on the effect on the momentum of the atom, when this is illuminated by a pulse of light, with a short time duration of the order of τ.

In the next section we will show that by suitably combining three pulses separated by longer time drifts, one theoretically proves the possibility of producing atom interferometry.

The first thing we have to do when a velocity $\boldsymbol{v}$ and consequently a momentum $\boldsymbol{p}$ of the atom enters the picture, is to write its Hamiltonian. Since the motion of the atom is essentially the motion of its barycentre, we will have

$$\boldsymbol{p}_b = m \cdot \boldsymbol{x}_b$$

and the dynamic system will be described in terms of the variables

$$\boldsymbol{x}_b \, , \; \boldsymbol{r} = \boldsymbol{x}_e - \boldsymbol{x}_b \, , \tag{6.60}$$

i.e. position of the barycentre and position of the electron with respect to the barycentre.

It is clear that $\boldsymbol{x}_b, \boldsymbol{r}$ refer to independent degrees of freedom and as such they commute as it happens for the respective moment. The Hamiltonian of the system now becomes

$$H = H_b + H_{0I} + \delta H \tag{6.61}$$

where

$$H_b = \frac{p_b^2}{2m}, \; . \tag{6.62}$$

H_{0I} is the internal Hamiltonian of the atom, that coincides with H_0 of the previous paragraph (see (6.28)), while

$$\delta H = -\boldsymbol{E} \cdot \boldsymbol{d} = -\boldsymbol{E} \cdot e\boldsymbol{r} \; . \tag{6.63}$$

The difference with respect to the treatment of Sect. 6.4, however, is that here in general the origin is not coinciding with the barycentre anymore, so that in describing the electrical field $\boldsymbol{E}$ one has to use the relation

$$\boldsymbol{E} = \boldsymbol{E}_0 \cos(\boldsymbol{k} \cdot \boldsymbol{x}_b - \omega t - \varphi) \; . \tag{6.64}$$

We have already seen that of the two frequencies $\pm\omega$ implied by (6.64), it is only $-\omega$ that generates a resonance with the internal degrees of freedom, so in (6.64) we can consider only the reduced field

$$\boldsymbol{E} = \frac{1}{2} E_0 e^{i(\boldsymbol{k} \cdot \boldsymbol{x}_b - \omega t - \varphi)} \; ; \tag{6.65}$$

this is the expression that will enter into our perturbation δH.

Now let us consider the free states, namely the eigenstates of

$$H_0 = \frac{p_b^2}{2m} + H_{0I} \,.$$

Here and in the subsequent formulas, no confusion should be made between the imaginary unit i, that is mostly in the exponentials, and the index i used to denote the free eigenstates.

Since p_b^2 and H_{0I} commute, we expect that the eigenstates can be factored into the product of the eigenstates of H_{0I}, namely u_{oi}, $(i = e, g)$ that we have already seen in Sect. 6.4, multiplied by

$$u_{ob} = e^{\pm i \frac{p_b \cdot x_b}{\hbar}} \,; \tag{6.66}$$

the free eigenstates have the form

$$w_{oi,\boldsymbol{p}} = u_{oi}(\boldsymbol{r}) e^{\pm i \frac{p_b \cdot x_b}{\hbar}} \,. \tag{6.67}$$

Since (6.66) represents two counter-propagating waves of matter, we will consider only one of them. For instance, when we consider an experiment of free fall, we can assume that $\boldsymbol{p} = m\dot{\boldsymbol{x}}_b$, as the velocity vector, is directed downward. So we restrict our attention to the eigenstates

$$w_{oi,\boldsymbol{p}} = u_{oi}(\boldsymbol{r}) e^{i \frac{\boldsymbol{p} \cdot \boldsymbol{x}_b}{\hbar}} \,, \quad (i = e, g) \,. \tag{6.68}$$

Note that we simplify the notation $\boldsymbol{p}_b$ to $\boldsymbol{p}$, since now there is no longer any ambiguity. The equation satisfied by $w_{oi,\boldsymbol{p}}$ is

$$(H_b + H_{0I}) w_{oi,\boldsymbol{p}} = \left(\frac{p^2}{2m} + E_{oi} \right) w_{oi,\boldsymbol{p}} \,. \tag{6.69}$$

For notation purposes, it is convenient to put

$$\omega_{i,p} = \frac{E_{oi,p}}{\hbar} = \frac{p^2}{2m\hbar} + \frac{E_{oi}}{\hbar} = \frac{p^2}{2m\hbar} + \omega_i \,. \tag{6.70}$$

Although $w_{oi,\boldsymbol{p}}$ is an improper eigenfunction in the variable $\boldsymbol{x}_b$, nevertheless two different states with $\boldsymbol{p}' \neq \boldsymbol{p}$ are orthogonal as established by Proposition 4.1. To avoid normalization problems, we could consider wave packets in $\boldsymbol{x}_b$, however we continue here by using only the plane wave with a single frequency for the sake of simplicity.

An important point we need to understand is that when the atom is falling (at the moment gravity does not enter into the picture) and the light is off, it proceeds by inertia with constant velocity, and it has a momentum $\boldsymbol{p}_b$ when it is in the ground state g, while it is augmented to $\boldsymbol{p}_b + \boldsymbol{q}$ for some $\boldsymbol{q}$ when the state is at the excited level e. The same happens when the light is flashed, with the only difference that when the interaction is working there can be a jump from g (with momentum $\boldsymbol{p}$) to e (with momentum $\boldsymbol{p}+\boldsymbol{q}$) in case of absorption of a quantum of light or, vice versa, in case of emission.

The conclusion is that of the four states

$$\begin{array}{ll}\text{(a) } w_{og,\boldsymbol{p}} & \text{(b) } w_{og,\boldsymbol{p}+\boldsymbol{q}} \\ \text{(c) } w_{oe,\boldsymbol{p}} & \text{(d) } w_{oe,\boldsymbol{p}+\boldsymbol{q}}\end{array}$$

only (a) and (d) are of physical interest.

This is important because we are going to simplify, as in Sect. 6.4, the analysis of the state of the system by restricting it to a linear combination of (a) and (d); namely, slightly simplifying the notation

$$\begin{cases} w_{og} = w_{og,\boldsymbol{p}} = u_{og} e^{i\frac{\boldsymbol{p}\cdot\boldsymbol{x}_b}{\hbar}} \\ w_{oe} = w_{oe,\boldsymbol{p}+\boldsymbol{q}} = u_{oe} e^{i\frac{(\boldsymbol{p}+\boldsymbol{q})\cdot\boldsymbol{x}_b}{\hbar}} \end{cases} \tag{6.71}$$

we will put

$$\psi(t) = a_e(t) w_{be} + a_g(t) w_{og} \,. \tag{6.72}$$

Remark 6.4 The fact that w_{oe}, w_{og} are orthogonal is a consequence of (4.59); in fact, considering that

$$< w_{oe}, w_{og} >=< e^{i\frac{\boldsymbol{p}\cdot\boldsymbol{x}_b}{\hbar}} \,,\, e^{i\frac{(\boldsymbol{p}+\boldsymbol{q})\cdot\boldsymbol{x}_b}{\hbar}} >< u_{oe}, u_{og} > \tag{6.73}$$

we see that both factors in (6.73) are zero, i.e. $w_{oe} \perp w_{og}$.

There is, however, a problem of normalization, because we know that an improper state cannot have a finite L^2 norm. This is typical of δ-normalization of improper states. However, we will rely on the following approximate orthogonality relation, that holds when the relevant wavelengths are much smaller than the volume V on which we integrate. Namely, consider the functions $q_k(x) = \left\{\frac{e^{-i2\pi \boldsymbol{k}\cdot x}}{\sqrt{V}}\right\}$; then we have

$$< q_k(x) \,,\, q_{k'}(x) >= \frac{1}{V}\int_V e^{i(\boldsymbol{k}-\boldsymbol{k}')\cdot x} dx \cong \begin{cases} 0 \; \boldsymbol{k} \neq \boldsymbol{k}' \\ 1 \; \boldsymbol{k} = \boldsymbol{k}' \,. \end{cases}$$

Based on that, since the normalization constant will not enter into our calculations, we will conventionally agree that

$$i = e, g \,,\, < w_{oi}, w_{oi} >=< e^{i\frac{\boldsymbol{p}\cdot\boldsymbol{x}}{\hbar}}, e^{i\frac{\boldsymbol{p}\cdot\boldsymbol{x}}{\hbar}} >< u_{oi}, u_{oi} >= 1 \,. \tag{6.74}$$

Note that for the moment we have not specified the value of $\boldsymbol{q}$ in (6.71) because we want to derive it from a consistent application of our quantum principles.

In this sense, since it is clear that the extra momentum $\boldsymbol{q}$ derives from the interaction with the e.m. field, we have to look at the action of δH on our stationary states $e.g.$

Going back to (6.63), (6.65) we see that δH can be written as

$$\begin{cases} \delta H = \delta H_0 e^{i\boldsymbol{k}\cdot\boldsymbol{x}_b}, \\ \delta H_0 = -\frac{1}{2}\boldsymbol{E}_0 \cdot e\,\boldsymbol{r} e^{-i(\omega t+\varphi)}\ , \end{cases} \tag{6.75}$$

where δH_0 is the same as that of Sect. 6.4 and acts only on functions of $\boldsymbol{r}$, while $e^{i\boldsymbol{k}\cdot\boldsymbol{x}_b}$ is a new factor acting only on functions of $\boldsymbol{x}_b$.

In particular we know that the action of δH on the state of the system is described by the cross-product $< w_{oe}, \delta H w_{og} >$ and $< w_{og}, \delta H w_{oe} >$ and, since the latter is just the conjugate of the former, we can consider only the former.

In view of (6.71) we have

$$\begin{aligned} \delta H w_{og} &= e^{i\boldsymbol{k}\cdot\boldsymbol{b}_b} \delta H_0 \left(u_{og} e^{i\frac{\boldsymbol{p}\cdot\boldsymbol{x}_b}{\hbar}} \right) \\ &= e^{i\frac{(\boldsymbol{p}+\hbar\boldsymbol{k})\cdot\boldsymbol{x}_b}{\hbar}} \delta H_0 u_{og}\ . \end{aligned} \tag{6.76}$$

Then we put

$$\boldsymbol{q} = \hbar\boldsymbol{k}\ . \tag{6.77}$$

We underline that, given the formula

$$|\boldsymbol{k}| = \frac{2\pi}{\lambda}\ ,$$

we know from general wave theory, (6.77) yields

$$|\boldsymbol{q}| = \frac{h}{2\pi}\frac{2\pi}{\lambda} = \frac{h}{\lambda}\ ,$$

namely we retrieve again equation (3.30) that expresses the momentum of a photon of wavelength λ, showing that the momentum conservation is also naturally embedded into quantum theory.

We now proceed exactly as in Sect. 6.4 and we derive the analogous of (6.31), namely

$$\begin{aligned} i\hbar(\dot{a}_e w_{oe} + \dot{a}_g w_{og}) = &\ \hbar(\omega_{e,|\boldsymbol{p}+\boldsymbol{q}|} a_e w_{oe} + \omega_{g,p} a_g w_{og}) + \\ &+ (a_e \delta H w_{oe} + a_g \delta H w_{og})\ . \end{aligned} \tag{6.78}$$

We assume again that δH cannot lead outside the space spanned by w_{oe}, w_{og} and transform (6.78) into

$$\begin{cases} i\dot{a}_e = \omega_{e,|\boldsymbol{p}+\boldsymbol{q}|}\ a_e + \frac{1}{\hbar} < w_{oe}, \delta H w_{oe} > a_e + \frac{1}{\hbar} < w_{oe}, \delta H w_{og} > a_g \\ i\dot{a}_g = \omega_{g,p}\ a_g + \frac{1}{\hbar} < w_{og}, \delta H w_{oe} > a_e + \frac{1}{\hbar} < w_{og}, \delta H w_{og} > a_g \ . \end{cases} \quad (6.79)$$

We first notice that again the diagonal elements in (6.79) are zero, i.e. recalling (6.22), (6.28) and (6.73),

$$\begin{cases} < w_{oe}, \delta H w_{oe} > = < e^{i\frac{(\boldsymbol{p}+\boldsymbol{q})\cdot \boldsymbol{x}_b}{\hbar}}, e^{i\boldsymbol{k}\cdot \boldsymbol{x}_b} e^{i\frac{(\boldsymbol{p}+\boldsymbol{q})\cdot \boldsymbol{x}_b}{\hbar}} > \cdot < u_{oe}, \delta H_0 u_{oe} >= 0 \\ < w_{og}, \delta H w_{og} > = < e^{i\frac{\boldsymbol{p}\cdot \boldsymbol{x}_b}{\hbar}}, e^{i\frac{(\boldsymbol{p}+\boldsymbol{q})\cdot \boldsymbol{x}_b}{\hbar}} >< u_{og}, \delta H_0 u_{og} >= 0 \ . \end{cases} \quad (6.80)$$

Then, recalling again (6.76), we observe that the off-diagonal element $\frac{1}{\hbar} < w_{oe}, \delta H w_{og} >$ can be written as

$$\begin{aligned} \frac{1}{\hbar} < w_{oe}, \delta H w_{og} > &= \frac{1}{\hbar} < e^{i\frac{(\boldsymbol{p}+\boldsymbol{q})\cdot \boldsymbol{x}_b}{\hbar}}, e^{i\frac{(\boldsymbol{p}+\boldsymbol{q})\cdot \boldsymbol{x}_b}{\hbar}} > \cdot < u_{oe}, \delta H_0 u_{og} >= \\ &= \frac{1}{\hbar} < u_{oe}, \delta H_0 u_{og} > \ . \end{aligned} \quad (6.81)$$

Adopting the convention of Sect. 6.4 (see (6.32), (6.35)), we call

$$\frac{1}{\hbar} < u_{oe}, \delta H_0 u_{og} >= \Omega \cos(\omega t + \varphi) \ , \quad (6.82)$$

so that (6.79) becomes

$$\begin{cases} i\dot{a}_e = \omega_{e,|\boldsymbol{p}+\boldsymbol{q}|}\ a_e + \Omega \cos(\omega t + \varphi) a_g \\ i\dot{a}_g = \omega_{g,p}\ a_g + \Omega^* \cos(\omega t + \varphi) a_e \end{cases} \quad (6.83)$$

which is formally identically to (6.38).

From here on we can follow the same route, namely we put

$$\begin{cases} a_e(t) = e^{-i\omega_{e,|\boldsymbol{p}+\boldsymbol{q}|}t} b_e(t) \\ a_g(t) = e^{-i\omega_{g,p}t} b_g(t), \end{cases} \quad (6.84)$$

$$\begin{cases} \Omega = \Gamma e^{i\alpha} \quad (\Gamma = |\Omega|) \\ \gamma = \varphi - \alpha, \end{cases} \quad (6.85)$$

$$\begin{aligned} \delta &= \omega - (\omega_{e,|\boldsymbol{p}+\boldsymbol{q}|)} - \omega_{g,p}) = \\ &= \omega - [\omega_e - \omega_g + \frac{1}{2m\hbar}(|\boldsymbol{p}+\boldsymbol{q}|^2 - p^2)] = \\ &= \omega - \left[\omega_e - \omega_g + \frac{\boldsymbol{p}\cdot\boldsymbol{q}}{m\hbar} + \frac{q^2}{2m\hbar}\right] . \end{aligned} \quad (6.86)$$

As we see, the major difference with respect to Sect. 6.4 is in the detuning (6.86), that now contains more terms than the previous $\delta = \omega - (\omega_e - \omega_g)$.

All this considered, we find the "slow" coefficients $b_e(t), b_g(t)$ to satisfy the system

$$\begin{cases} i\dot{b}_e = \frac{1}{2}\Gamma e^{-i(\delta t+\gamma)} b_g \\ i\dot{b}_g = \frac{1}{2}\Gamma e^{i(\delta t+\gamma)} b_e \,, \end{cases} \tag{6.87}$$

the solution of which is given by (6.51) with the same definitions of Ω_r, $\sin\vartheta$, $\cos\vartheta$ as in (6.46).

Once again we remark that, when the e.m. field is off, we have $\Gamma = 0$ and hence $b_e(t), b_g(t)$ remain constant in time.

Summarizing we can say that the result of this section is that introducing the velocity and the relative momentum of the barycentre, we find formally that the slow coefficients have the same form as in Sect. 6.4, with the difference that the detuning is now given by (6.86) and that the stationary states between which the atom can swap are now $w_{oe,|\boldsymbol{p}+\boldsymbol{q}|}$, $w_{og,p}$.

6.5 The Atom Interferometry and the Measurement of *g*

At first we will prove that by a combination of flashes of light and periods of pure drifts, an atom, possessing a momentum $\boldsymbol{p}$ and a detuning δ given by (6.86), can produce an interferometry, namely a mixture of states g and e that does depend on the duration of flashes and drifts. Then we will introduce the gravity acceleration, proving that the probability of the two states depends on $|\boldsymbol{g}| = g$, which, therefore, becomes measurable. No confusion should arise between $|\boldsymbol{g}|$ and the index g.

Indeed, when the experiment is simultaneously conducted on a cloud of $\sim 10^6$ or more atoms, the probability will be transformed into a percentage, that we will be able to count by their fluorescence effect.

To fix the ideas and in view of the target of measuring g, we will consider an atom just falling along the vertical and a laser beam propagating in the same direction.

Consider the following sequence of time intervals (see Fig. 6.1): a first period τ of interaction, a longer time drift T, a second period of interaction 2τ, a long time drift T, a final period of interaction τ.

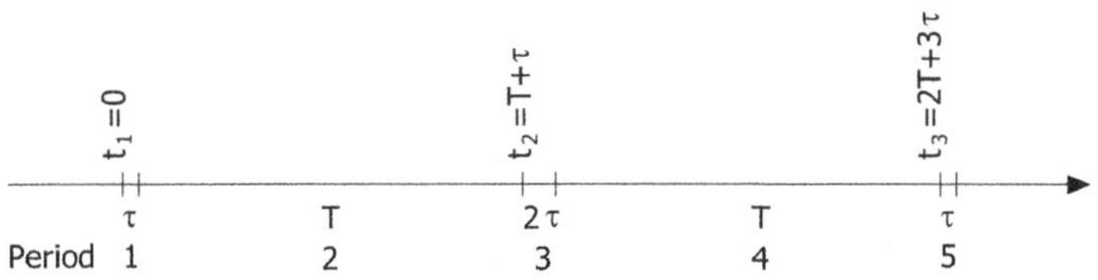

Fig. 6.1 The different time phases of the interaction light–atom

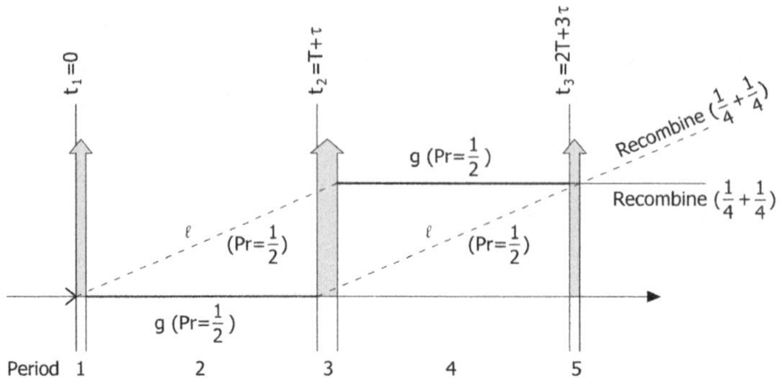

Fig. 6.2 The scheme of the three-pulses atom interferometry, $\Omega_r \tau = \pi/2$, without gravity

The value of τ is chosen in such a way that

$$\Omega_r \tau = \frac{\pi}{2}\ , \quad \Omega_r 2\tau = \pi\ . \tag{6.88}$$

While τ is very small (in fact for the experiment that employs Cesium atoms we typically have $\Omega_r \sim 6 \times 10^3 \mathrm{s}^{-1}$ and $\tau \sim 2.5 \times 10^{-4}$s), T is longer and variable between 0.1 s and 1 s.

According to Examples 6.1 and 6.2, the first pulse acts as a splitter (with probabilities $\frac{1}{2}, \frac{1}{2}$) between the two stationary states, the second pulse acts as a mirror, the third pulse again as a splitter, that, however, here has the opposite function of recombining the states. In this sense, there is a deep analogy between the actual phenomenon of atom interferometry and the optical interferometry of Mach–Zehnder type.

We follow now the evolution of "slow" coefficients through the periods 1, 2, 3, 4, 5 of Fig. 6.2, assuming that at the beginning

$$b_e(0) = 0\ , \quad b_g(0) = 1\ . \tag{6.89}$$

Then repeatedly using (6.51), (6.56), (6.58), considering that every pulse of light has its own initial phase φ, and therefore its value of $\gamma = \varphi - \alpha$, and assuming that we are in a condition of quasi-resonance, so that $\sin\vartheta \cong 1$, $\cos\vartheta \cong 0$, we find

$$\text{period 1} \begin{cases} b_e(\tau) = \dfrac{e^{-i\frac{\delta\tau}{2}}}{\sqrt{2}}(-ie^{-i\gamma_1}) \\ b_g(\tau) = \dfrac{e^{i\frac{\delta\tau}{2}}}{\sqrt{2}}\ , \end{cases} \tag{6.90}$$

$$\text{period 2} \begin{cases} b_e(t_2) = b_e(\tau) \\ b_g(t_2) = b_g(\tau)\ , \end{cases} \tag{6.91}$$

$$\text{period 3} \begin{cases} b_e(t_2 + 2\tau) = e^{-i\delta\tau}(-ie^{-i\gamma_2})b_g(t_2) \\ b_g(t_2 + 2\tau) = e^{i\delta\tau}(-ie^{i\gamma_2})b_e(t_2)\ , \end{cases} \tag{6.92}$$

$$\text{period 4} \begin{cases} b_e(t_3) = b_e(t_2 + 2\tau) \\ b_g(t_3) = b_g(t_2 + 2\tau) \ , \end{cases} \tag{6.93}$$

$$\text{period 5} \begin{cases} b_e(t_3 + \tau) = e^{-i\delta\frac{\tau}{2}} \left[\frac{1}{\sqrt{2}} b_e(t_3) - i\frac{1}{\sqrt{2}} e^{-i\gamma_3} b_g(t_3)\right] \\ b_g(t_3 + \tau) = e^{i\delta\frac{\tau}{2}} \left[-i\frac{1}{\sqrt{2}} e^{i\gamma_3} b_e(t_3) + \frac{1}{\sqrt{2}} b_g(t_3)\right] . \end{cases} \tag{6.94}$$

Substituting one into the other we obtain for instance, for $b_e(t_3 + \tau)$

$$b_e(t_3 + \tau) = \frac{1}{2} e^{-i\delta\tau} \left[-1 + e^{i(\delta\tau + 2\gamma_2 - \gamma_1 - \gamma_3)}\right] ; \tag{6.95}$$

therefore, setting

$$\Delta\varphi = \gamma_1 - 2\gamma_2 + \gamma_3 = \varphi_1 - 2\varphi_2 + \varphi_3 \ , \tag{6.96}$$

we get

$$|b_e(t_3 + \tau)|^2 = \frac{1}{2}[1 - \cos(\delta\tau - \Delta\varphi)] . \tag{6.97}$$

Equation (6.97) gives the probability that the final state of the system is in e.

We can now introduce the action of gravity, which indeed has the effect of changing linearly in time (chirp) the term $p = mgt$ in (6.86).

Therefore, the detuning δ now reads, recalling that $q = \hbar k$,

$$\delta = \omega - (\omega_e - \omega_g + gtk + \frac{\hbar}{2m} k^2) . \tag{6.98}$$

Let us notice that (6.98) is written in terms of purely scalar quantities; this is because we have assumed that $\boldsymbol{k}, \boldsymbol{q}$ and $\boldsymbol{p}$ all have the same direction and verse. We remark as well that, while the "Doppler term" gtk is of the order of 2π MHz after a time of 0.1 s (namely the span, T, of the periods 2 and 4 in Fig. 6.1), the last constant term of (6.98), for a Cesium atom is of the order of 2π kHz, namely three orders of magnitude smaller than the former and, as such, negligible.

If we like to take the detuning δ close to 0 in each of the periods 1, 3, 5 in which light pulses are active, we have to increase the frequency ω of the laser beam. This has been done in the first interferometric experiments [2] by using simultaneously three signal generators with frequencies, respectively

$$\omega_1 = \omega_0 \ , \ \omega_2 = \omega_0 + \omega_m \ , \ \omega_3 = \omega_0 + 2\omega_m \ ,$$

in such a way that, if ω_0 were resonant at $t = 0$ with $\omega_e - \omega_g$, after the time $t_2 \sim T$ (here we can use the approximation $\tau = 0$), the new detuning δ_2 would be $\delta_2 \cong \delta_1 = 0$, namely

$$\omega_m = gTk \ ,$$

i.e., the second signal becomes resonant with the atom that at the beginning of period 3 has acquired a velocity gT. Then again after a long period T, namely at time $t_3 \cong 2T$, the third signal will be resonant with the atom that has acquired a velocity $g \cdot 2T$ implying an increase of frequency

$$2\omega_m = g \cdot 2T \cdot k\ .$$

Given that, the phase of each signal propagates in space–time, when the radiation meets the atom at time t and coordinate $z = z_0 - \frac{1}{2}gt^2$, as

$$\varphi_m(t) = \omega_m t - k(z_0 - \frac{1}{2}gt^2)\ , \tag{6.99}$$

where z_0 is the height from which the atom has been dropped; since this is arbitrary we prefer to fix $z_0 = 0$.

If we write (6.99) for each n at time t_n, namely ($t_1 \cong 0\ ,\ t_2 \cong T\ ,\ t_3 \cong 2T$)

$$\begin{cases} \varphi(t_1) = 0 \\ \varphi_2(t_2) = \omega_0 T + \omega_m T + \frac{1}{2}kgT^2 \\ \varphi(t_3) = \omega_0 2T + 2\omega_m 2T + \frac{1}{2}kg(2T)^2\ . \end{cases} \tag{6.100}$$

This gives the interferometric phase-shift

$$\Delta\varphi = \varphi_3 - 2\varphi_2 + \varphi_1 = 2\omega_m T + kgT^2\ , \tag{6.101}$$

so that from (6.97) we ultimately derive

$$P_r(\text{system in } e \text{ at } 2T) = \frac{1}{2}[1 - \cos(2\omega_m T + kgT^2)]\ . \tag{6.102}$$

In other words, by measuring the fraction of atoms finally in the state e we can come to know kgT^2.

By changing T one can have different values of the probability (6.102), that translate into different percentages over the population of a certain quantity of atoms and then derive g from he observed phase-shift $2\omega_m T + kgT^2$ by a least squares fit.

The above reasoning is based on theoretical ground, however in order to reach an accuracy of a very few μGal, a number of practical improvements need to be implemented.

6.6 Complements and Improvements

The formulas derived in Sect. 6.5 are based on the assumption that the atoms in a gas cloud have an initial velocity equal to zero. If they all possessed the same initial velocity, along the z-axis, then the calculations would not be very different. However, we have to take into account that the thermal energy chaotically spreads the velocity of atoms according to Boltzmann law

$$\sigma(v) = \sqrt{< v^2 >} = \sqrt{\frac{3k_B T_e}{m}}$$

with k_B the Boltzmann constant, T_e the absolute temperature and m the mass of the atom. So, if at room temperature we have $\sigma(v) \sim 10^3 \mathrm{ms}^{-1}$, at 10 nKelvin the velocity is reduced to less than 6 mm s^{-1}. Therefore, one understands that the availability of a technique to cool atoms down to temperatures of a few nKelvin, has been a fundamental ingredient, without which no reasonable measurement of gravity could be possible. In the literature (e.g. see [3, 4]), temperatures as low as 5 nKelvin are reported.

It is interesting to remark that such cooling is again the result of a fine manipulation of the atom gas by laser beams.

We cannot enter here in this field which has nowadays many applications beyond the problem of gravity measurement. For such a subject the reader can consult the literature, for instance, [5].

For what concerns our matter, it is fundamental to state that a sample of atoms useful for the measurement of g has first to be prepared in a cold state at the level of the nano Kelvin.

A second point of practical interest is that, like in classical gravimeters, measurement operations can be done on a sample of atoms first launched in the upwards direction interfering with the two branches of the motion, ascending and descending.

A fundamental point to improve quite significantly the accuracy of the result is to perform a two-photon stimulated Raman transition. Essentially the idea is that increasing the momentum transfer factor k one can improve the sensitivity of the measurement and so its accuracy too.

On the other hand, the two states g, e have to be stable enough, in terms of time, to allow a measurement with a sufficiently long T. For this reason, the states are chosen in the hyperfine structure that is created by different spin orientations of the electrons in ground state (see Remark 5.3). The point is, however, that the energy difference of the two states is by definition very small and so is the value of k for the laser beam in resonance with them, necessary to produce the Raman effect.

The seaming contradiction is solved by introducing a two-photon stimulated Raman transition. Basically one introduces an "intermediary" energy level i, much higher than g and e as in Fig. 6.3.

The level i is generally not stable, but this is not a problem because what we are looking for is an "instantaneous" transition $g \to i \to e$ or vice versa.

In order to achieve this result, one has to use two laser beams, one ascending and the other descending, with frequencies ω_1 and ω_2 such that

$$\begin{cases} \omega_{ge} = \omega_e - \omega_g = \omega_1 - \omega_2 \\ \omega_1 = \omega_i - \omega_e \, , \; \omega_2 = \omega_i - \omega_g \, . \end{cases} \tag{6.103}$$

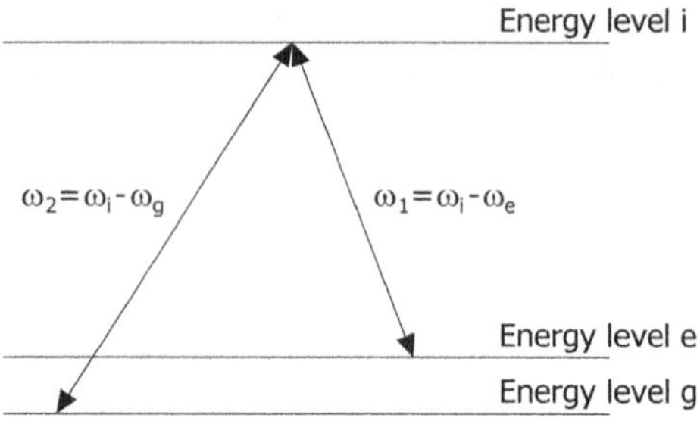

Fig. 6.3 The scheme of stimulated two-photon Raman transition, with perfect resonance

Notice that, as suggested by Fig. 6.3, we have

$$\omega_{ge} \ll \omega_1, \omega_2 \ . \tag{6.104}$$

The first laser beam is capable of stimulating the transition $g \to i$, or the reverse, while the second stimulates "immediately" the transition $i \to e$, or the reverse. The final result in terms of energy is a jump

$$E_{ge} = \hbar(\omega_1 - \omega_2) = \hbar\omega_{ge} \tag{6.105}$$

exactly as the one we studied in the previous section. However the Doppler term $\delta\omega_D$ in the detuning becomes much different, in fact the momentum conservation imposes that the total momentum transfer is $\hbar\boldsymbol{k}_1 - \hbar\boldsymbol{k}_2$ with $\hbar\boldsymbol{k}_1 = \boldsymbol{q}_1$ referring to the absorption of a photon in the jump $g \to i$, while $\hbar\boldsymbol{k}_2 = \boldsymbol{q}_2$ refers to the emission of a photon in the jump $i \to e$. The point is that $\boldsymbol{k}_2$ has a direction opposed to $\boldsymbol{k}_1$ because it is stimulated by the second beam travelling in a direction contrary to that of the first. Since the modulus of $\boldsymbol{k}_1, \boldsymbol{k}_2$ are fixed by ω_1, ω_2,

$$k_1 = \frac{\omega_1}{c} \ , \ k_2 = \frac{\omega_2}{c}$$

and since $\omega_1 \cong \omega_2$, because of (6.104), we have $k_1 \cong k_2$ too. Therefore, the Doppler term of the two-photon transition is

$$\delta\omega_D = \frac{p \cdot (\boldsymbol{q}_1 - \boldsymbol{q}_2)}{m\hbar} = \frac{\boldsymbol{p} \cdot (\boldsymbol{k}_1 - \boldsymbol{k}_2)}{m} \equiv \frac{\boldsymbol{p} \cdot \boldsymbol{k}_{eff}}{m} \tag{6.106}$$

with

$$\boldsymbol{k}_{eff} = \boldsymbol{k}_1 - \boldsymbol{k}_2 \ , \ k_{eff} \cong 2k_1 \ . \tag{6.107}$$

One can analytically prove that then the evolution of the dynamic system essentially follows the lines of Sect. 6.6, on condition that we use $\boldsymbol{k}_{eff}$ instead of $\boldsymbol{k}$ and the implied modification of the other parameters of the dynamics (see [6]).

The result of this solution, with a measurement based on Cesium atoms, is to multiply the k factor by 10^5, which is a decisive tool for accurate measurement of g. In fact, the accuracy is also five orders of magnitude better.

Still another tool can be introduced into the measurement process, making it more precise. In fact, by combining in a short sequence of impulses with a π or a $\pi/2$ phase shift, but obtained with beams propagating in alternating directions, with each impulse one can transmit an impulse $\hbar k_{eff}$ to the atom, so improving by an integer factor the term k_{eff} (when a two-photon stimulated Raman transition is used) in the Doppler term that provides the measurement of g.

6.7 Results and Perspectives

Measuring g by cooled atoms interferometry is a very hot matter, with results improving every year. Here we shall refer first of all to a not so recent paper (see [7]), where the target of 3μGal (3×10^{-9} g) precision has been achieved, although in 2008 [8] reported an experiment where an uncertainty of 1.3×10^{-9} g is ascertained.

The reason why we refer to an older experiment is that not only the uncertainty is determined but also the observations from the atom interferometer are compared to those of a falling corner-cube FG-5 instrument, over a time span of three days. In the experiment, conducted at the Laboratory of the Physics Department of Stanford University, a cloud of 5×10^8 atoms was created and first cooled at the level of 1.5 μK. Then the cloud was vertically launched with a velocity of 3 ms^{-1}. During launch the temperature in the z direction was further lowered to the level of 10 nK; this permitted the creation of interferometric fringes determined by Raman transitions, stimulated by two counter-propagating laser beams, with a sequence of impulses with T varying between 0.5 and 160 ms.

The number of atoms taking part into the interferometric process was not that of all the atoms launched, yet it was of the order of 3×10^6, so giving a quite large statistics as for the result. In Fig. 6.5 we report the fringes from [7]. Each of the 40 points of the figure represents a single launch of the atoms, spanning a period of 1 minute. The least squares fit of the curve gives an estimate of the accuracy of $\sim$ 3μGal (Fig. 6.4).

Another quite interesting result of the experiment is shown in Fig. 6.5, where three days of data from the atom interferometer are compared with the FG-5 data.

As we see, the spread of the former is about three times smaller than that of the latter. The difference of the absolute level of the gravity between the two is essentially due to a difference of height between the two instruments. The final discrepancy between the two is $\sim$ 7μGal, attributed by the authors mainly to the uncertainty of the vertical gravity gradient.

Factors that can induce systematic errors have been also analysed [9]. Among them, those with a possible influence above 1μGal, but below 10μGal, are: phase-shifts of Raman frequency pulses, knowledge or Coriolis force and of the gravity gradient, presence of synchronous noise, knowledge of pressure, tidal and other

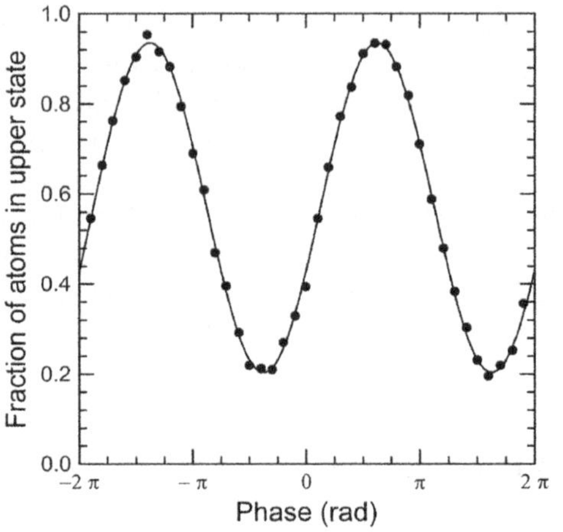

Fig. 6.4 Fringes of the interferometry experiment at Stanford 1997. Peters A., K. Y. Chung and S. Chu High-precision gravity measurements using atom interferometry, *Metrologia* 38 25, 2001

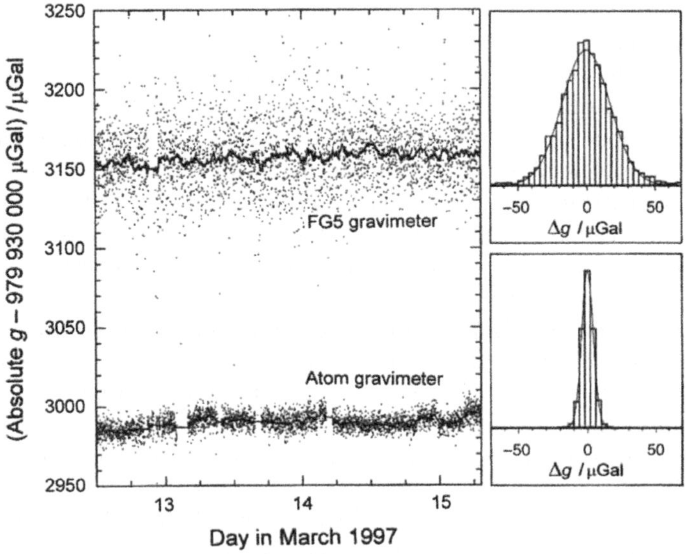

Fig. 6.5 Comparison of the results of atom interferometry with FG-5 measurements in the Stanford experiment. Peters A., K. Y. Chung and S. Chu High-precision gravity measurements using atom interferometry, *Metrologia* 38 25, 2001

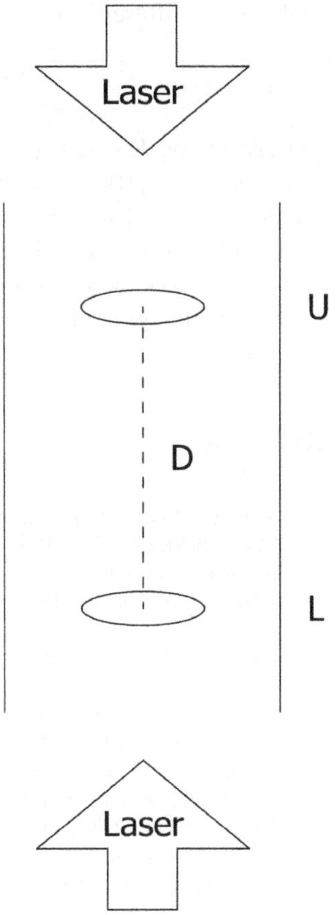

Fig. 6.6 Arrangement of an atom interferometry gradiometer; the arrows represent counter-propagating laser beams producing the Raman effect; U, L, upper and lower atom clouds; D distance between U and L

environmental corrections. Some other systematic effects have been reduced in recent years below the μGal level and, at the same time, even those mentioned above have been reduced by an order of magnitude.

Therefore, considering technical improvements, we could say that a standard target for present-day atom interferometry is the 1 μGal accuracy level, while in the next future we expect to go to an even higher accuracy.

Before closing, we would like to mention that the same technique described in these notes can be used to perform gradiometric measurements too.

The simplest configuration is represented schematically in Fig. 6.6 (see [10]).

The phase of the fringes of the two clouds are given by (see (6.102))

$$\varphi_U = 2\omega_m T + k_{eff} g_U T^2$$
$$\varphi_L = 2\omega_m T + k_{eff} g_L T^2$$

so that their difference is

$$\Delta\varphi = k_{eff} T^2 (g_U - g_L)$$

that divided by D (see Fig. 6.6) gives the vertical gradient of gravity.

In literature [10], experiments on ground report an accuracy of ~ 10 E, for a distance D of ~ 1 m, where 1 E (1 Eötvos) = 10^{-9}s^{-2}

It is interesting to note however that would the experiment be portable to a satellite in free fall, since the integration time T could be expanded by a factor 10^3 up to 10 s, in principle the gradient of g could be determined at the level of 1 mE.

References

1. Berman PR (Ed) (1997) Atom interferometry, Academic Press Inc
2. Kasevich M, Chu S (1992) Measurement of the gravitational acceleration of an atom with a light-pulse atom interferometer. Appl Phys B 54:321. https://doi.org/10.1007/BF00325375
3. Cohen-Tannoudji CN (1998) Manipulating atoms with photons. Rev Modern Phys 70(3):707–719
4. Phillips WD (1998) Laser cooling and trapping of neutral atoms. Rev Modern Phys 70(3):721–741
5. Bordé CJ (1997) Matter wave interferometry: a synthetic approach. In: Atom Interferometry, Academic Press Inc, pp 257–290
6. Young B, Kasevich M, Chu S (1997) Precision atom interferometry with light pulses. In: Atom interferometry, Academic Press Inc, pp 363–405
7. Peters A, Chung KY, Chu S (2001) High-precision gravity measurements using atom interferometry. Metrologia 38:25–61
8. Müller H, Chiow S, Herrmann S, Chu S, Chung KY (2008) Atom interferometry tests of the isotropy of post-Newtonian gravity. Phys Rev Lett 100:031101
9. de Angelis M, Bertoldi A, Cacciapuoti L, Giorgini A, Lamporesi G, Prevedelli M, Saccorotti G, Sorrentino F, Tino GM (2009) Precision gravimetry with atomic sensors. Meas Sci Tech 20:022001
10. Tino GM, Sorrentino F, Aguilera D, Battelier B, Bertoldi A, Bodart Q, Bongs K, Bouyer P, Braxmaier C, Cacciapuoti L, Gaaloul N, Gurlebeck N, Hauth M, Herrmann S, Krutzik M, Kubelka A, Landragin A, Milke A, Peters A, Rasel EM, Rocco E, Schubert C, Schuldt T, Sengstock K, Wicht A (2013) Precision gravity tests with atom interferometry in space. Nuclear Phys B Proc Suppl 243244:203217

Further Reading

11. Landau LD, Lifšic EM (1962) Mécanique quantique. Boringhieri, Torino
12. Landau LD, Lifšic EM (1965) Meccanica. Boringhieri, Torino
13. Landau LD, Lifšic EM (1966) Physique théorique. Tome 3, Mécanique quantique. Théorie non relativiste, Ed Mir Moscou
14. de Broglie L (1951) Onde, corpuscoli, meccanica ondulatoria. Bompiani, Modena
15. Heisenberg W (1930) The physical principles of the quantum theory. The University of Chicago Press, Chicago, Translated by C. Eckart and F. C Hoyt

16. Caldirola P, Cirelli R, Prosperi GM (1982) Introduzione alla fisica teorica. UTET, Torino
17. Chu S, Cohen-Tannoudji CN, Phillips W (1998) Nobel Lecture 1997. Rev Mod Phys 70(30):685–741
18. Pauli W (1962) Meccanica ondulatoria. Boringhieri, Torino
19. Pauli W (1964) Elettrodinamica. Boringhieri, Torino
20. Messiah A (1965) Mécanique quantique 1–2. Dunod, Paris

The manufacturer's authorised representative in the EU is Springer Nature Customer Service Centre GmbH, Europaplatz 3, 69115 Heidelberg, Germany. If you have any concerns regarding our products, please contact ProductSafety@springernature.com

Printed and bound by CPI Group (UK) Ltd, Croydon, CR0 4YY
30/09/2026
02225470-0008